EARLY PLIOCENE MARINE CLIMATE AND ENVIRONMENT OF THE EASTERN VENTURA BASIN, SOUTHERN CALIFORNIA

EARLY PLIOCENE MARINE CLIMATE AND ENVIRONMENT OF THE EASTERN VENTURA BASIN, SOUTHERN CALIFORNIA

BY

J. PHILIP KERN

UNIVERSITY OF CALIFORNIA PRESS
BERKELEY • LOS ANGELES • LONDON
1973

UNIVERSITY OF CALIFORNIA PUBLICATIONS IN GEOLOGICAL SCIENCES

Approved for publication September 15, 1971
Issued March 30, 1973

UNIVERSITY OF CALIFORNIA PRESS
BERKELEY AND LOS ANGELES
CALIFORNIA

◇

UNIVERSITY OF CALIFORNIA PRESS, LTD.
LONDON, ENGLAND

ISBN: 0-520-09424-7

LIBRARY OF CONGRESS CATALOG CARD NUMBER: 72–182549

PRINTED IN THE UNITED STATES OF AMERICA

To Claudia

CONTENTS

EARLY PLIOCENE MARINE CLIMATE AND ENVIRONMENT OF THE EASTERN VENTURA BASIN, SOUTHERN CALIFORNIA

BY
J. PHILIP KERN

ABSTRACT

About 300 feet of conglomerate, sandstone, and siltstone of the Towsley Formation crop out in Elsmere and Grapevine canyons at the eastern end of the Ventura basin. The abundant and diverse fossil fauna in these rocks includes 129 species of molluscs, 3 species of echinoderms, and 3 species of sharks. An early Pliocene age is suggested by known stratigraphic ranges of 23 diagnostic taxa, including *Lyropecten estrellanus, Astrodapsis fernandoensis, Chione elsmerensis, Nassarius hamlini, Patinopecten lohri, Mytilus coalingensis, Dendraster,* and *Turritella cooperi,* and by comparison with the fauna of the regional lower Pliocene type section in the San Joaquin Valley and with other lower Pliocene faunas in central California. This paper records the first comprehensive study of the Towsley Formation fauna, the only large shallow-water fauna of early Pliocene age in southern California.

The Elsmere Canyon fauna suggests deposition in 10 to 30 fathoms of water, probably within a mile or less of an exposed shore. Stratigraphic faunal changes suggest gradual shoaling of the water during deposition. Orientation of elongate shells occurred locally through the action of weak currents from the northeast or the southwest, but these currents were not strong enough to cause significant postmortem transport of shells. Beds of shells probably were formed by changes in rates of deposition of sediments.

The present-day distribution of the Recent species and of the nearest Recent relatives of the extinct species is used in this study to interpret the early Pliocene marine climate and environment of deposition. The geographic ranges of all but 5 of the 67 Recent species overlap in a narrow zone about one degree of latitude south of Elsmere Canyon, suggesting that there has been a slight northward shift since early Pliocene time in the distribution of these species and that early Pliocene marine climate in the Ventura basin was slightly warmer than it is today.

This interpretation is opposed to the conclusions of previous studies, most of which have postulated substantially warmer early Pliocene marine climate in this region. The earlier interpretations are based on the presence in most west coast lower Pliocene faunas of molluscan taxa that today are restricted to warmer southern waters and on the assumptions that distribution is limited by temperature effects on survival and that northern range end points are determined by minimum winter temperature. Some species are, however, apparently limited by temperature effects on reproduction or on other stages of the life cycle, and their northern range end points are determined by maximum summer temperature, duration of the warm season, or other thermal variables. Distribution of other species may be independent of temperature control, and some taxa clearly have evolved physiologically and have adapted to different thermal conditions.

The Elsmere Canyon fossil fauna contains 14 southern molluscan taxa that do not live close to this latitude today, suggesting some sort of climatic change since early Pliocene time. The distribution of these taxa in the Gulf of California and on the outer coast of Lower California suggests that their northern range end points may be determined by maximum summer temperature or duration of the warm season rather than by minimum survival temperature. The influence of the California Current on early Pliocene summer nearshore temperature was probably different from today because of the different configuration of the coastline. The restriction of cold south-flowing water out of the deeply embayed coastal basins, possibly in combination with solar heating of these shallow embayments, may have resulted in slightly higher summer temperatures, possibly accounting for the presence of a few southern molluscan taxa north of their Recent range end points in southern and central California. The higher summer temperatures may have been associated with an annual thermal regime only slightly warmer (perhaps 1°C) than at present, accounting for the slight northward shift of Recent species since early Pliocene time.

INTRODUCTION

Nature and Scope of the Investigation

Shallow-water marine sedimentary rocks of the lower Pliocene Towsley Formation crop out in a small area in Elsmere and Grapevine canyons at the eastern end of the Ventura basin. These rocks contain an abundant and diverse megafauna (hereafter referred to as the Elsmere Canyon fauna), the only large shallow-water assemblage of this age in southern California. The paleoclimatic implications of the Elsmere Canyon fauna have been evaluated in a number of previous studies (see below), but the fossil assemblage has never been thoroughly documented. This study was undertaken to reconstruct the environment in which the Elsmere Canyon fauna lived, by comparison with environments in which the same and similar species live today. Answers have been sought to the following specific questions:

1) What is the age of the Elsmere Canyon fauna?

2) What was the nature of the physical environment in which the fauna lived? What was the topographic form of the depositional area? What was the degree of exposure to the open sea? How deep was the water?

3) What was the nature of the marine thermal regime in which the fauna lived?

Geologic Setting

Rocks of the Towsley Formation were deposited in the Ventura basin during late Miocene and early Pliocene time. This study is concerned with the fossil fauna from the middle part of the formation where it is exposed in Elsmere and Grapevine canyons, which drain the west end of the San Gabriel Mountains east of Newhall, Los Angeles County, California (figs. 1, 2). The Ventura basin extends from this area westward for more than 100 miles, the western half forming the present-day Santa Barbara Channel. It is bounded on the north by the Santa Ynez and Topatopa mountains and on the south by the Santa Monica Mountains and their westward extension, the Channel Islands. This marine basin was one of a number that existed along the California coast during late Tertiary time (Reed and Hollister, 1936:13–51). Basins that are known to have received marine sediments during late Miocene and early Pliocene time are shown in figure 1.

The southern California coastal region was submerged during much of early and middle Tertiary time (Corey, 1954: figs. 2–6). This condition persisted until late in the Miocene Epoch, when marine water covered the present coastal area from Los Angeles to Castaic to Santa Maria (Reed, 1933*a:* fig. 42; Edwards, 1934:805; Corey, 1954: fig. 7). At the close of the Miocene Epoch the Ventura basin was isolated from the Santa Maria basin in the north and the Los Angeles basin in the south by uplift of the Santa Ynez, San Gabriel, and Santa Monica mountains (Natland and Kuenen, 1951:83). The degree of separation of the Los Angeles and Ventura basins is a subject of controversy. The paleogeographic maps of Edwards (1934: figs. 3–5) and Corey (1954: fig. 8) show two completely

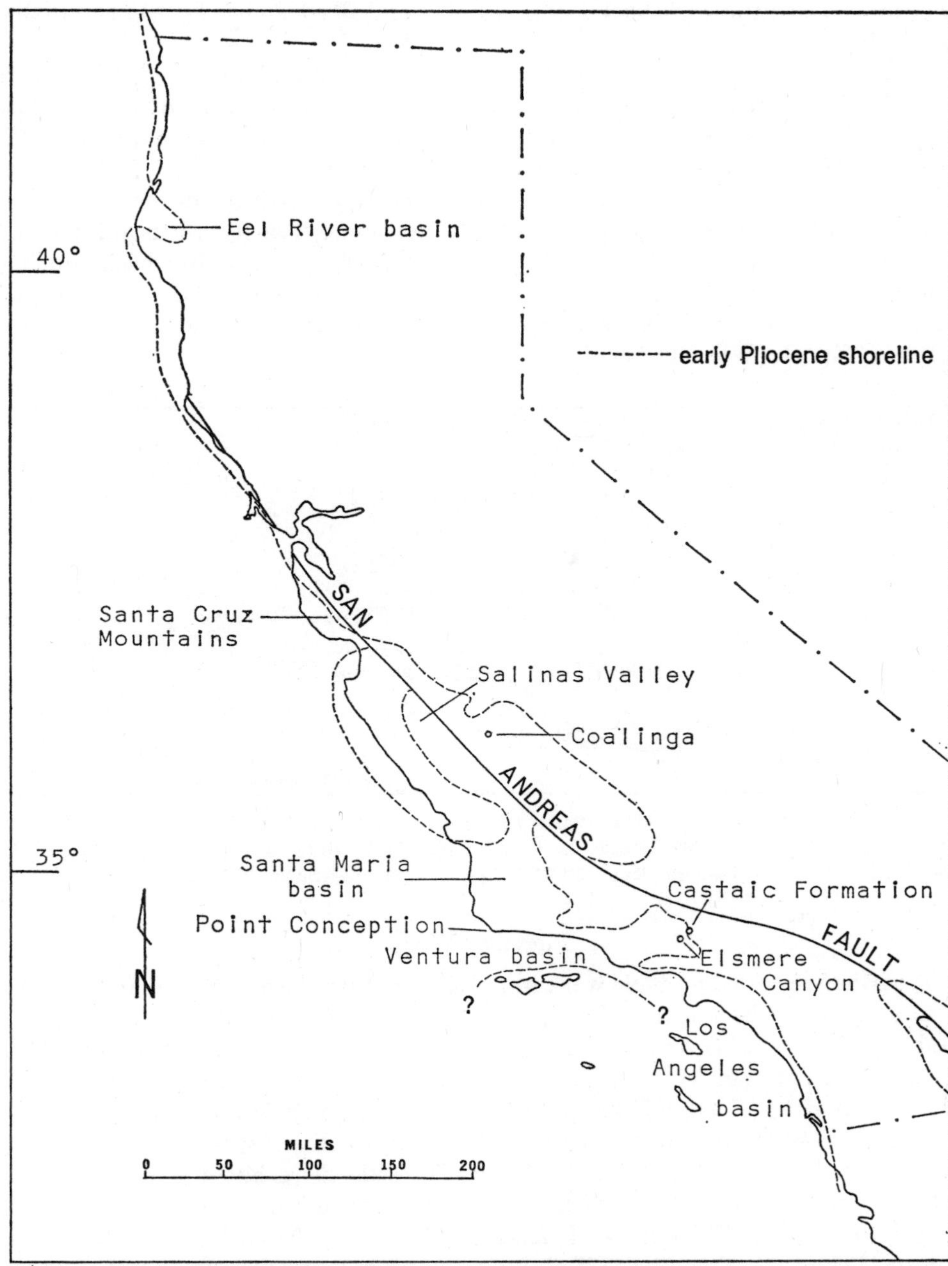

Fig. 1. Localities of lower Pliocene fossil faunas and inferred early Pliocene shoreline. Sources of paleogeographic data: Reed, 1933*a*: fig. 51; Reed and Hollister, 1936; Corey, 1954: fig. 8; Axelrod, 1956: fig. 16.

separate basins; those of Clark (1921: fig. 11) and Reed 1933*a:* fig. 51) show only one large basin (fig. 3).

These Pliocene basins were once thought to have been shallow (Reed, 1933*a:*3–5; 1933*b* :231), mainly because of the predominance of coarse-grained clastic rocks. Foraminiferal faunas, however, suggest that the Ventura basin was at least 4,000 to 5,000 feet deep during early Pliocene time (Natland, 1957:566–567), when rocks of the Towsley Formation now exposed in Elsmere and Grapevine canyons were being deposited in shallow nearshore environments. Twenty thousand feet of marine sediments were deposited in the center of the subsiding basin during Pliocene and early Pleistocene time (Natland and Kuenen, 1951:83). The Los Angeles basin had a similar history, shoaling from 4,000 to 900 feet during the Pliocene Epoch (Woodford et al., 1954:73).

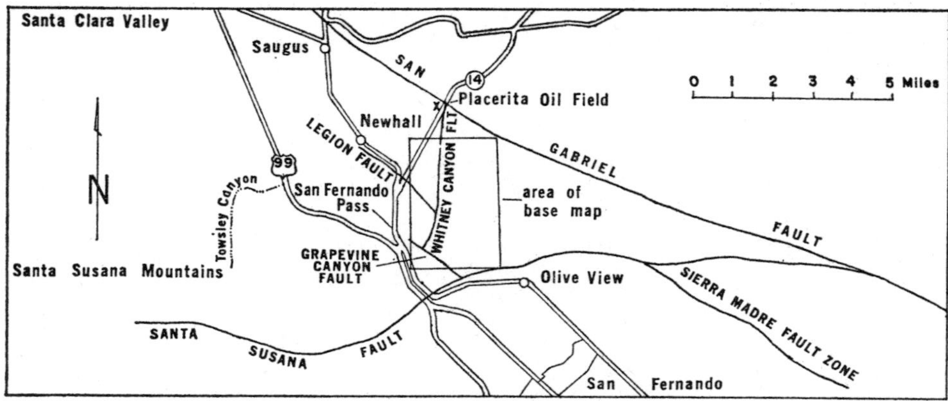

Fig. 2. Index map and regional structural features in the area of Elsmere and Grapevine canyons. Sources of structural data: Hill, 1937: pl. 15; Winterer and Durham, 1962: pl. 44; Calif. Div. of Mines Geologic Map of California, Los Angeles Sheet, 1955 preliminary edition.

Previous Work

The earliest recorded geologic work in the area of Elsmere and Grapevine canyons was done by members of the Geological Survey of California (Whitney, 1865:121). Gabb (1866–1869:49) described Pliocene fossils collected by the survey near San Fernando Pass (fig. 2). Later fossil collections were made from the same rock units at the site of the San Fernando tunnel by Ashley (1895:337–339).

The first reference to the Elsmere Canyon localities was made by Watts (1901:56), who collected "middle Neocene" fossils there but did not publish a faunal list. Lists of fossils from Elsmere Canyon have been published by Arnold (1907*a:*525–527), Eldridge and Arnold (1907:25), English (1914:209–211), Smith (1919:152), Kew (1924:77–79), Carson (1926), and Grant and Gale (1931).

The Elsmere Canyon area was included in early geologic maps of the eastern Ventura basin by Eldridge and Arnold (1907: pl. 1) and by Kew (1924: pl. 1). More recently these rocks have been mapped by Oakeshott (1950), who surveyed the geology of the San Fernando quadrangle, and by Willis (1952), who correlated them with oil-producing units in the Placerita oil field (fig. 2). Winterer and

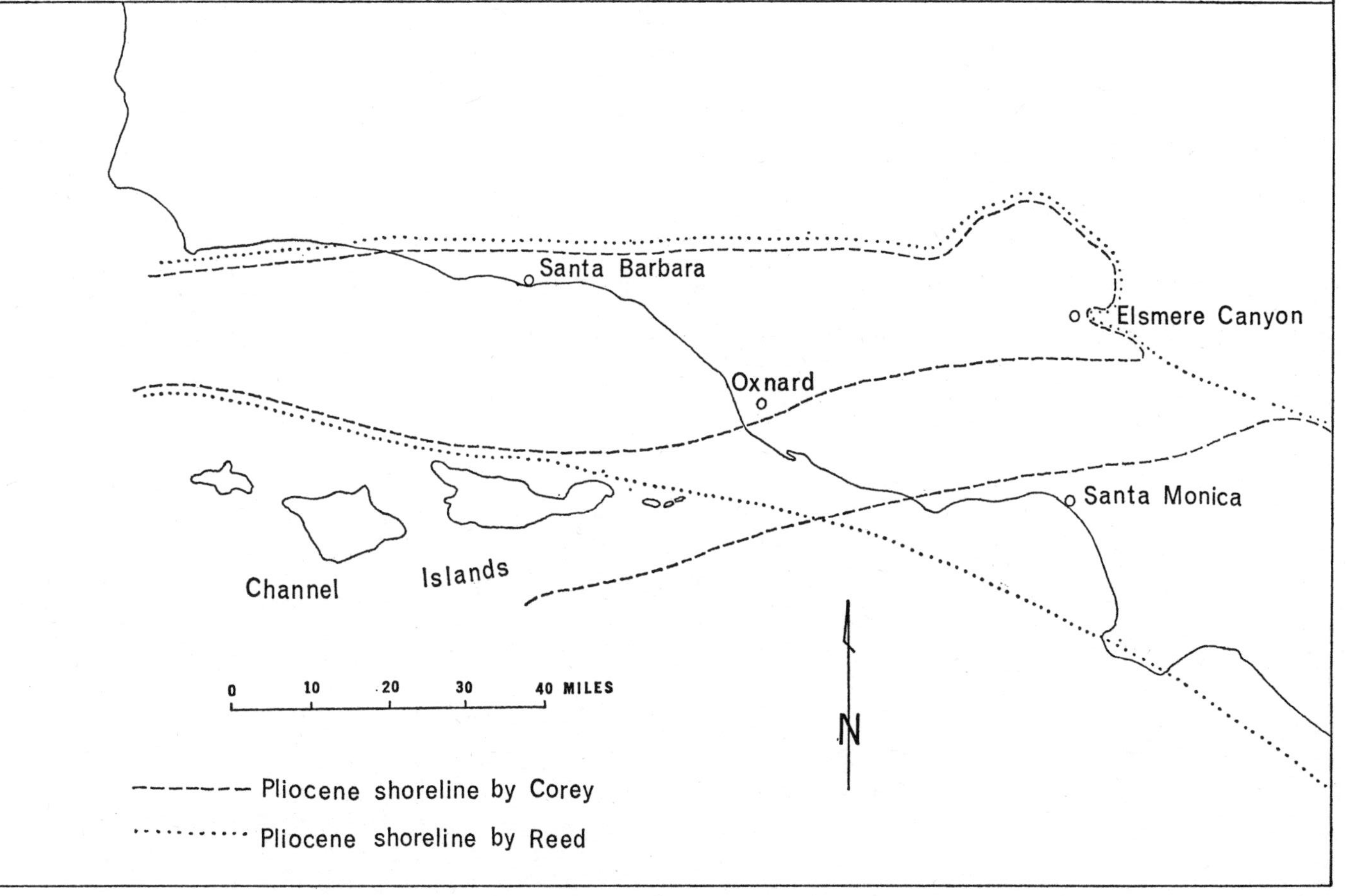

Fig. 3. Pliocene paleogeography of southern California, from Reed (1933*a*: fig. 51) and Corey (1954: fig. 8).

Durham (1951:2631; 1954: map sheet 5; 1962:287–308) named and described the Towsley Formation, including the Elsmere Canyon section.

Fieldwork

My fieldwork was done from December 1966 to May 1967, and from October to November 1967. Mapping was done on the U.S. Geological Survey 7.5-minute San Fernando quadrangle enlarged to a scale of 1:4,800.

Acknowledgments

Dr. Clarence Hall of the University of California, Los Angeles, deserves much of the credit for this paper. Dr. Hall first brought to my attention the significance of the Elsmere Canyon fauna; he generously gave of his time and knowledge throughout the course of the study and he painstakingly read the entire manuscript in various stages of completion and through several stages of evolution of ideas. I am very grateful to Dr. Hall for his contribution to this study.

The penultimate draft of the manuscript benefited from the suggestions of members of my dissertation committee at the University of California, Los Angeles: Professors N. Gary Lane and Gerhard Oertel of the Department of Geology, Leonard Muscatine and Peter P. Vaughn of the Department of Zoology, and Preston E. Cloud, now of the University of California, Santa Barbara. The present paper has been revised substantially from that draft, and countless improvements have been made in the final paper because of the meticulous work of Professor James W. Valentine, University of California, Davis, and Warren O. Addicott, U.S. Geological Survey, who reviewed the manuscript, and of Mrs. Grace H. Stimson of the University of California Press. To all these people I am most grateful.

Numerous other individuals provided assistance with various aspects of the study. Dr. John E. Warme, now of Rice University, has been very helpful in innumerable ways. Dr. D. I. Axelrod of the University of California, Davis, kindly discussed with me his views on late Tertiary climate. Mrs. LouElla Saul of the University of California, Los Angeles, helped identify some of the fossil molluscs. The fossil sharks were identified by Dr. Shelton Applegate of the Los Angeles County Museum. I was accompanied in the field at different times by Dr. Larry Frakes, now at Florida State University, John Grimmer and Irv Neder of the University of California, Los Angeles, and Sigurd Heiberg, now at the Norwegian Geotechnical Institute; their advice on stratigraphic and paleontologic problems is gratefully acknowledged. The preparation of illustrations for this report was facilitated by the advice and suggestions of Mrs. Opal Kurtz, John Grimmer, and Dave Weide, all of the University of California, Los Angeles.

Financial support has been provided by a National Defense Education Act Title IV Fellowship. Permission to work in the Elsmere Canyon area was granted by the Ventura office of the Standard Oil Company of California.

STRATIGRAPHY

Tertiary sedimentary rocks overlie pre-Tertiary igneous and metamorphic rocks at the eastern end of the Ventura basin. In the area of Elsmere and Grapevine canyons the Tertiary section includes an unnamed Eocene unit and the Pliocene Towsley and Pico formations. The youngest rocks in this area are Quaternary terrace and landslide deposits.

Pre-Tertiary Rocks

The igneous and metamorphic rocks underlying the Tertiary formations on the western flank of the San Gabriel Mountains have been described by Oakeshott (1958:50–56). In Elsmere and Grapevine canyons these rocks include schist, quartzite, and marble of the Placerita Formation, granodiorite, and intrusive diorite gneiss. Clasts of all these rock types are common in conglomerates in the Tertiary formations.

Eocene Series

The mapped area is divided into two structural blocks by the Whitney Canyon fault (fig. 4, in pocket). Pre-Tertiary rocks are overlain on the west side of this fault by sandstone and siltstone of Eocene age. More than 6,000 feet of Eocene rocks are penetrated by wells in Elsmere Canyon, and though it is possible that the section has been repeated by faulting, Winterer and Durham (1962:337) suggested a minimum original thickness of at least 1,300 feet a short distance west of the fault. In the subsurface of the Placerita oil field, two miles north of Elsmere Canyon (fig. 2), this unit is more than 2,000 feet thick. It has not been recognized anywhere east of the Whitney Canyon fault. In Elsmere Canyon the Pliocene Towsley Formation unconformably overlies the Eocene rocks on the west side of the fault. Distribution, lithology, and evidence for the age of these rocks have been discussed by Winterer and Durham (1962:282–283).

Pliocene Series

TOWSLEY FORMATION

Distribution.—The type section of the Towsley Formation is in Towsley Canyon on the north slope of the Santa Susana Mountains (fig. 2). The formation is 4,000 feet thick in this area, where the north limb of the Pico anticline plunges beneath the Santa Clara Valley (Winterer and Durham, 1962:288). In the Elsmere Canyon area the middle part of the Towsley Formation overlaps the Miocene Modelo Formation to lie on Eocene rocks and on the pre-Tertiary basement, and it is overlapped in turn by the Pico Formation. Lithology of the Towsley Formation west of the Elsmere Canyon area has been described in detail by Winterer and Durham (1962:289–292).

Nomenclature.—Eldridge and Arnold (1907:23–25, pl. 1) mapped all the rocks between the Miocene Modelo and Pleistocene Saugus formations in the eastern Ventura basin as the Pliocene Fernando Formation. They included in this unit fossiliferous rocks in lower Elsmere Canyon but mapped underlying beds in upper Elsmere Canyon as the Miocene Vaqueros Formation. (See fig. 5 for comparison with present treatment.)

Kew (1924:70–72, pl. 1) renamed all post-Modelo, pre-Saugus rocks in the Ventura basin the Pico Formation. He also recognized the twofold subdivision of the rocks in Elsmere Canyon, including the lower unit in his Pico Formation and the overlying rocks in the Saugus Formation (fig. 5).

<table>
<tr><th>Eldridge and Anrold, 1907</th><th colspan="2">Kew, 1924</th><th>Oakeshott, 1950</th><th colspan="2">Willis, 1952</th><th>Winterer and Durham, 1962
this report</th></tr>
<tr><td>Fernando Formation
(middle Pliocene)</td><td rowspan="2">Fernando Group</td><td>Saugus Formation</td><td>Pico Formation</td><td rowspan="2">Pico Formation</td><td></td><td>Pico Formation</td></tr>
<tr><td>Vaqueros Formation
(lower Miocene)</td><td>Pico Formation</td><td>Repetto Formation</td><td>"Repetto equivalent"</td><td>Towsley Formation</td></tr>
<tr><td colspan="7">OLDER ROCKS</td></tr>
</table>

Fig. 5. Names applied to the two Pliocene formations in Elsmere and Grapevine canyons by previous workers and in the present study. The older rocks in this area include pre-Tertiary igneous and metamorphic rocks and Eocene sedimentary rocks.

Oakeshott (1950:50–51, table 1) called the upper unit the Pico Formation and the lower unit the Elsmere Member of the Repetto Formation, correlating these rocks with lower Pliocene rocks of the Los Angeles basin. Both units were included in the Pico Formation by Willis (1952:32), who referred to the lower one as "Repetto and Repetto equivalent" (fig. 5).

The Ventura and Los Angeles basins, however, were probably isolated from each other during Pliocene time, and they contain lithologically different rocks. Since the type section of the Repetto Formation is in the Los Angeles basin, the name "Repetto" should not be used in the Ventura basin. Consequently Winterer and Durham (1954, map sheet 5) proposed the name "Towsley Formation" for the sequence of siltstone, sandstone, and conglomerate between the Modelo and Pico formations in the Santa Susana Mountains. In the area west of Elsmere Canyon these rocks had been included in the Modelo Formation by Kew (1924: pl. 1). Winterer and Durham (1962: pl. 44) mapped the lower strata in Elsmere Canyon as part of the Towsley Formation and included the overlying rocks in the Pico Formation. This treatment has been followed in the present study (fig. 5).

Subdivision of the Towsley Formation in Elsmere Canyon.—The Towsley Formation in Elsmere and Grapevine canyons is about 300 feet thick. I have subdivided this part of the formation into three stratigraphic units, informally designated the lower unit, the upper unit, and the sandstone-conglomerate tongue of the upper unit (fig. 6, in pocket). In the Placerita oil field these units are known, respectively, as the lower Kraft zone, the lower Kraft shale, and the upper sand and conglomerate (Willis, 1952:32, 35, 36). Because these units apparently are of only local utility, formal stratigraphic names are not proposed for them.

Lower unit.—The thickness of the lower unit ranges from 100 feet in Grapevine Canyon to 50 feet east of the Grapevine-Elsmere divide to 175 feet in upper Elsmere Canyon (fig. 6, in pocket). West of the Whitney Canyon fault this unit is 100 feet thick. It varies in thickness from 200 to 950 feet in the subsurface of the Placerita oil field (Willis, 1952:38).

In the mapped area the Towsley Formation is everywhere unconformable on older rocks, west of the Whitney Canyon fault on Eocene rocks and east of the fault on pre-Tertiary rocks. Deposition apparently was initiated on a surface of very low relief, and throughout the mapped area this contact is nearly flat with local relief of only a few feet. Where the Pliocene rocks have been eroded from the slopes east of Elsmere Canyon, a remarkably planar pre-Towsley erosion surface is exposed.

The lower unit consists of siltstone, sandstone, and conglomerate, and locally the basal few feet contain large angular blocks derived from the immediately underlying rocks. The most extensive of several lenticular bodies of basal conglomerate has a maximum thickness of 25 to 30 feet and a lateral extent of about 1,000 feet. Bedding is irregular and poorly defined. The matrix of the conglomerate is coarse-grained, pebbly sandstone, of the same composition as the basal sandstone between the conglomerate bodies. Most clasts in the conglomerate are well rounded and range in size from 2 to 10 inches.

Most of the lower unit is composed of siltstone, fine sandy siltstone, and silty, fine-grained to coarse-grained sandstone. These rock types grade into one another. In general both the percentage of sand and the size of the sand grains are larger near the base of the unit. Sandstone is restricted to the basal few feet in Grapevine Canyon, where most of the unit consists of siltstone, but in upper Elsmere Canyon the basal sandstone is 75 feet thick. The silty coarse-grained basal sandstone in upper Elsmere Canyon is bituminous; it is dark brown to black on fresh surfaces but weathers to light brown or gray. This rock is soft, though slightly more resistant than the finer-grained rocks higher in the section, which contain no bitumen and are lighter-colored. There is no evident bedding in the lower unit except where concretionary layers are developed locally.

Calcium carbonate is present as a cementing agent in some beds in the lower unit. The fossil-bearing beds and several unfossiliferous beds in upper Elsmere Canyon are so cemented. Most of these beds are cemented irregularly and discontinuously, and many of them consist only of planar alignments of concretions. The cement probably is all of concretionary origin. Concentrations of fossils apparently have served as nuclei around which calcium carbonate was deposited to form resistant beds and nodules. It may be suggested that shells were originally abundant throughout, and that only those in calcareous beds have been preserved. Beds of well-preserved shells, however, commonly pass unchanged through pockets of uncemented rock, and well-preserved shells occur locally outside the cemented beds, indicating that fossil shells have not been leached from the uncemented rock.

Calcium carbonate also serves as a cementing agent in brecciated fault zones cutting the lower unit. Origin of the cement in fault zones or in concretionary beds is not known. It may have been derived in part from fossil shells, especially in fault zones, or in part from marble in the pre-Tertiary metamorphic rocks. Bodies of cemented rock occur isolated within bituminous rock where the two occur together, suggesting that mineralization preceded immigration of the bitumen.

The bitumen content of these rocks apparently is proportional to the percentage and grain size of sand. The lower unit is one of the oil-producing zones (the lower Kraft zone) in the Placerita oil field (Willis, 1952:38) and probably in the Elsmere Canyon field. Asphalt-like bitumen has seeped up through fractures at several places along the north–south faults in Grapevine Canyon, and it flows with water from springs along the Whitney Canyon fault in Elsmere Canyon. There are a number of prospect pits in the thick basal bituminous sandstone in upper Elsmere Canyon.

Charcoal fragments are disseminated throughout the lower unit and are locally concentrated in thin beds. Similar occurrences were observed by Natland and Kuenen (1951:101) in the Towsley Formation in the central part of the Ventura basin. Shepard (1951:56) found charcoal fragments in sand layers in some of the deep basins of the continental borderland off southern California. These fragments, and those in the Towsley Formation, are not compressed; they are irregularly shaped, not resembling plant fragments; and they occur commonly with bits of noncarbonized wood. Thus they were not carbonized after deposition as plant fragments but were deposited as charcoal. Thin layers of charcoal commonly accumulate on southern California beaches during winter storms after summers of large brush fires in the mountains. These layers seem a likely source for the charcoal in the Towsley Formation, as well as for that in the Recent deep basins.

Isolated fossil-bearing beds and lenses are distributed irregularly throughout the lower unit in Grapevine Canyon and the southern half of the Elsmere Canyon drainage. Seven fossiliferous horizons are exposed in the steep northwest wall of upper Elsmere Canyon (fig. 6, in pocket). Several of these beds crop out continuously for as much as 1,500 feet (fig. 4, in pocket). Fossils are also abundant in the basal conglomerate in the southeast corners of sections 7 and 18 (fig. 4). The distribution and abundance of fossils by locality are shown in figure 9 (in pocket).

Upper unit.—The thickness of the upper unit is variable (fig. 6). It is 300 feet thick in the subsurface south of the mapped area (Winterer and Durham, 1962:293). East of the Whitney Canyon fault the thickness decreases northward from 230 feet in Grapevine Canyon to about 200 feet at the Grapevine-Elsmere divide and 150 feet in upper Elsmere Canyon. To the north this unit is overlapped by the Pico Formation. West of the fault it is about 250 feet thick south of Elsmere Canyon and 100 to 150 feet thick in the Placerita oil field (Willis, 1952:38), where it is known as the lower Kraft shale.

The contact with the lower unit is not everywhere well defined, though it represents at least a local break in deposition, and the siltstone of the upper unit was deposited on an irregular, channeled surface at two places in the mapped area. The upper unit truncates the highest beds of the lower unit in upper Elsmere Canyon, and concretionary beds in the lower unit are truncated east of UCLA locality 5778 (fig. 4, in pocket) by the overlying siltstone. The siltstone of the upper unit is overlain in part by the sandstone-conglomerate tongue and in part by conglomerate and sandstone of the Pico Formation.

In Grapevine and upper Elsmere canyons the upper unit is composed of light brown siltstone and fine sandy siltstone (the sandstone-conglomerate tongue is described separately). This rock is less well consolidated than the siltstone and sandstone of the lower unit. It weathers readily, typically forming gentle slopes covered with tan silt. A distinctive feature of this unit is the local development of a network of intersecting dark brown limonitic veins. These veins are generally perpendicular to bedding, are from 0.5 inch to 1.5 inches thick, and apparently differ from the intervening rock only in having a high limonite content. No carbonate is present, but gypsum is locally abundant. The limonitic veins are not present everywhere in this unit but occur locally throughout the mapped area. They were not observed in the lower unit.

In lower Elsmere Canyon the siltstone of the upper unit grades laterally into sandstone and conglomerate. Since there are few outcrops in this area, the relationships between different rock types are not clear, but Winterer and Durham (1962:293) recognized the lateral gradation of Towsley siltstone into Pico sandstone and conglomerate. Thus the contact between these units is gradational and transgresses time, and the base of the Pico Formation is stratigraphically lower toward the west.

Concretions occur locally in the upper unit, and several fossiliferous beds in Grapevine Canyon are cemented. Very finely disseminated charcoal locally makes up as much as 1 percent of this unit. Bitumen has not penetrated the siltstone except along faults; the fine grain size may account for the impermeability of this rock to bitumen. Fossils occur above the basal few feet of the upper unit at only four localities.

Sandstone-conglomerate tongue of upper unit.—A tongue of sandstone and conglomerate in the upper unit thickens from 20 feet in Grapevine Canyon to 50 feet in upper Elsmere Canyon (fig. 6, in pocket). In the Placerita oil field the upper sandstone and conglomerate is 50 to 200 feet thick and lies directly under the Pico Formation and over siltstone of the upper unit (Willis, 1952:32). The interbedded sandstone and conglomerate interfingers with coarser-grained Pico conglomerate in upper Elsmere Canyon. In Grapevine Canyon these rocks are within the upper unit and are overlain by 10 to 30 feet of siltstone. They wedge out toward the east on the Grapevine-Elsmere divide, and west of the Whitney Canyon fault they interfinger with siltstone of the upper unit. To the west, in lower Elsmere Canyon, these rocks are mapped as undifferentiated siltstone, sandstone, and conglomerate, and they probably grade laterally into the Pico Formation.

The medium- to coarse-grained pebbly sandstone is poorly sorted and commonly grades into conglomerate. Clasts are well rounded and range in size from 1 to 4 or 5 inches. Large-scale cross-beds and channels indicate rapid deposition by strong currents. Fossils are rare in this unit but have been collected at two localities in siltstone lenses (UCLA localities 5780 and 5805). In Elsmere Canyon the rock is not bituminous, though it is part of an oil-producing zone (the upper sand and conglomerate) in the Placerita oil field (Willis, 1952:38).

Fossils.—The Elsmere Canyon fauna includes at least 129 species of molluscs, 3 species of echinoderms, and 3 species of sharks. All but 8 of the molluscan species have been collected during the present study. Three of these 8 species are in the invertebrate paleontology collection at the University of California, Los Angeles; the other 5 have been recorded by previous workers. Of the 135 species in the fauna, 109 have been definitely identified; the other 26 species are doubtfully identified or are referred only to genus. The entire known fauna is listed in figure 9 (in pocket) and in the Systematic Paleontology section.

Age and correlation of the Towsley Formation in Elsmere Canyon.—Because of incomplete knowledge of invertebrate megafossil faunas, the lack of thick, continuously fossiliferous stratigraphic sections, and environmental control of benthic invertebrate species, the Tertiary marine megafossil chronology on the west coast of North America is not so useful as the microfossil chronology. The historical development of the megafossil chronology used in this region has been summarized by Durham and Addicott (1965:A16) and by Adegoke (1969:65–69). The difficulties of correlating these regional time-stratigraphic units with the stages of the European type sections have been emphasized by Weaver et al. (1944:574), Durham (1954:23), Addicott (1969*b*:74), and others.

Grant and Gale (1931) employed a threefold time-stratigraphic subdivision of Pliocene strata in California. Their lower, middle, and upper Pliocene units were referred to the "Jacalitos zone," "San Diego zone," and "Santa Barbara zone," respectively (pp. 69–70). Woodring, Stewart, and Richards (1940:104–106) similarly referred to the Jacalitos, Etchegoin, and San Joaquin formations in the San Joaquin Valley as lower, middle, and upper Pliocene, respectively. This classification corresponds to Grant and Gale's (1931) treatment of these three formations. The threefold subdivision was formalized by Weaver et al. (1944: chart 11), whose time-stratigraphic classification of west coast Cenozoic rocks has generally been accepted as the regional standard. In this classification the lower and middle Pliocene "stages" are based, respectively, on the Jacalitos and Etchegoin formations.

Recently it has been recommended that a twofold subdivision of Pliocene strata would be more suitable (Vedder, 1960:B327; Durham and Addicott, 1965:A16–A17; Addicott, 1969*b*:74–75) because of the short duration of the Pliocene Epoch. I have retained the threefold subdivision, however, because of its wide acceptance and continued use (for example, Adegoke, 1969) and because I think the Elsmere Canyon fauna can be dated in such a classification.

In most previous studies of the Elsmere Canyon section of the Towsley Formation these rocks have been assigned an early Pliocene age. Yet it has long been recognized that a small component of the fauna is of Miocene aspect. For example, Grant and Gale (1931:30) listed as more characteristic of the Miocene than of the Pliocene, "*Pecten estrellanus* (typical variety?) and the variety *catalinae* but not variety *cerrosensis*, and *Ficus (Trophosycon) ocoyana* varieties, as well as the echinoid genus *Astrodapsis*, represented by *A. fernandoensis* Pack." Also, Durham (1948:1386) stated that "except for *Astrodapsis fernandoensis*, *Dendraster* sp. (*excentricus* auct.), and *Patinopecten lohri* all the important species of the Elsmere Canyon fauna have been found in strata at the base of the Modelo(?)."

In comparing the Towsley and Castaic formations, however, Stanton (1966:24) concluded that the upper Miocene Castaic fauna "resembles the Elsmere Canyon fauna but seems to be slightly older; in this paper, the Elsmere Canyon fauna is considered early Pliocene." Winterer and Durham (1962:321) placed the Miocene-Pliocene boundary "in the interval from about 500 to 1,000 feet above the base of the Towsley Formation at Towsley Canyon." Though precise correlation is difficult because of structural complexities, Winterer and Durham (1962:320) estimated that the basal beds in Elsmere Canyon are about 1,500 feet above the base at Towsley Canyon, that is, that they are Pliocene in age. These interpretations have recently been supported by Addicott (1970*b*:296, table 2), who also suggested an early Pliocene age for the Elsmere Canyon fauna.

Figure 7 graphically summarizes the stratigraphic ranges of 41 taxa that are critical in determining the age of the Elsmere Canyon fauna. (The ranges of all the species in the fauna and the sources of data are given in the Systematic Paleontology section.) Only one species has not been reported previously from rocks younger than Miocene: *Nassarius whitneyi* is known only from middle and upper Miocene rocks of California (Addicott, 1965:10). *N. whitneyi* occurs in Elsmere Canyon only in the basal beds. On the other hand, of the 103 definitely identified species in the Elsmere Canyon fauna, 41 are restricted to rocks younger

SPECIES	Upper Miocene	Lower Pliocene	Middle Pliocene	Upper Pliocene	Pleistocene	Recent
Nassarius whitneyi	← —					
Lyropecten estrellanus	← —	—				
Maxwellia eldridgei	—	—				
Nassarius stocki	—	—				
Astrodapsis fernandoensis	—	—				
Chione elsmerensis	—	—				
Nassarius hamlini		—				
Dosinia jacalitosana		—				
Calyptraea filosa	← —	—	—			
Ficus ocoyana	← —	—	—			
Astraea gradata	—	—	—			
Clathrodrillia coalingensis	—	—	—			
Calicantharus fortis angulatus	—	—	—			
Barbatia pseudoillota		—	—			
Patinopecten lohri		—	—			
Periploma planiuscula			—	—	—	—
Diodora murina			—	—	—	—
Nassarius iniquus		—	—	—		
Mytilus coalingensis		—	—	—		
Cancellaria arnoldi		—	—	—		
Calicantharus humerosus		—	—	—	—	
Dendraster		—	—	—	—	—
Tegula gallina		—	—	—	—	—
Turritella cooperi		—	—	—	—	—
Crepidula nummaria		—	—	—	—	—
Forreria belcheri		—	—	—	—	—
Ocenebra subangulata		—	—	—	—	—
Kelletia kelletii		—	—	—	—	—
Mitra idae		—	—	—	—	—
Megasurcula carpenteriana		—	—	—	—	—
Pododesmus cepio		—	—	—	—	—
Lima hemphilli		—	—	—	—	—
Lucina californica		—	—	—	—	—
Spisula hemphilli		—	—	—	—	—
Macoma yoldiformis		—	—	—	—	—
Gari californica		—	—	—	—	—
Callithaca tenerrima		—	—	—	—	—
Tivela stultorum		—	—	—	—	—
Dentalium neohexagonum		—	—	—	—	—
Haliotis fulgens		—	—	—	—	—
Fusinus barbarensis		—	—	—	—	—

Fig. 7. Stratigraphic ranges of species that are critical to determination of the age of the Elsmere Canyon fauna. Arrows indicate ranges that extend into pre–upper Miocene rocks. Sources of data are given in Systematic Paleontology section.

than Miocene. These include the diagnostic post-Miocene taxa *Patinopecten lohri, Mytilus coalingensis, Dendraster,* and *Turritella cooperi.* These taxa clearly indicate a post-Miocene age for the fauna.

Eleven of the post-Miocene species are known elsewhere only from upper Pliocene or younger deposits (table 1). Some of these species are small and obscure enough to have escaped detection in fossil faunas; some may have been misidentified in previous studies. Their occurrence here is regarded as extending their known stratigraphic ranges rather than as indicating a younger age for these rocks.

TABLE 1

ELSMERE CANYON MOLLUSCAN SPECIES NOT PREVIOUSLY REPORTED FROM ROCKS OLDER THAN LATE PLIOCENE

Species	Previously known range
Mangelia interlirata	Late Pliocene to Recent
Margarites pupillus	Late Pliocene to Recent
Taranis incultus	Late Pliocene to Recent
Acteon punctocaelatus	Early Pleistocene to Recent
Admete rhyssa	Early Pleistocene to Recent
Ocenebra foveolata	Pleistocene to Recent
Calliostoma splendens	Pleistocene to Recent
Balcis rutila	Pleistocene to Recent
Glycymeris profunda	Pleistocene to Recent
Amphissa reticulata	Recent
Ophiodermella cancellata	Recent

Two species have been recorded previously only from middle Pliocene and younger rocks (fig. 7). *Diodora murina* previously was known from Pliocene rocks only in the Los Angeles basin (Grant and Gale, 1931:850). Since a complete lower Pliocene section is not exposed in this basin and the fossil fauna is not well known (Woodring, 1938:11; Woodring, Bramlette, and Kew, 1946:42), this species may have been living there in early Pliocene time. *Periploma planiscula* has been recorded in Pliocene rocks only from the Etchegoin Formation near Coalinga (Grant and Gale, 1931:255, 854); apparently it is not common there, as it was not reported by Adegoke (1969).

On the other hand seven species, not including *Nassarius whitneyi,* have been recorded previously only from lower Pliocene and older rocks (fig. 7). These include the diagnostic pre–middle Pliocene species *Lyropecten estrellanus, Astrodapsis fernandoensis,* and *Chione elsmerensis,* and two species—*Nassarius hamlini* and *Dosinia jacalitosana*—previously reported only from lower Pliocene rocks. Thus the known stratigraphic ranges of individual taxa suggest that the Elsmere Canyon fauna is of early Pliocene age.

The "Jacalitos" Formation of Arnold (1909) is the generally accepted regional lower Pliocene standard in the west coast megafaunal chronology, as explained above. Nomland (1917) redefined Arnold's Etchegoin Formation to include both the Jacalitos and the overlying Etchegoin. Adegoke (1969:26–31, 43–48) has pointed out the absence of lithologic differences between Arnold's "Jacalitos" and lower "Etchegoin" and has combined these two units in the Etchegoin Formation. Thus the lower Etchegoin Formation is now the lower Pliocene standard in the west coast marine megafaunal chronology.

Correlation of the Elsmere Canyon fauna with that of the lower Etchegoin (Jacalitos) has been suggested by Grant and Gale (1931:30, 61, table 1), Winterer and Durham (1962:320), and Adegoke (1969:46). This correlation is predicated on the fact that the two faunas have about half of their species in common. Adegoke (1969:48) assigned the entire Etchegoin Formation a Pliocene age and correlated it with the upper Plaisancian and lower Astian stages of the standard European section. He distinguished two faunizones including five zonules in the Etchegoin Formation (pp. 26–31, 43–48). The Elsmere Canyon fauna is most similar to that of Zonule 10 in Faunizone F, the middle one in the fivefold subdivision. Five of the species whose lowest occurrences are in that zonule and three whose highest occurrences are in it also occur in Elsmere Canyon. Some ambiguity attaches to this designation in that one Elsmere Canyon species does not occur higher than Zonule 9 at Coalinga and one other Elsmere Canyon species does not occur lower than Zonule 12. The stratigraphic position of Zonule 10 relative to Arnold's formations is not clear, but apparently it is high in his "Jacalitos" Formation or low in his "Etchegoin." Unfortunately Adegoke did not define precise relationships between his individual zonules and the standard regional time scale, but his Zonule 10 appears to indicate an age of early or middle Pliocene. Thus an early Pliocene age for the Elsmere Canyon fauna is suggested by the stratigraphic ranges of individual species, by comparison with the Castaic Formation, and by correlation with the main body of the Towsley Formation and with the regional standard lower Pliocene section in the San Joaquin Valley.

Twelve of the 19 definitely identified mollusc and echinoid species in the shallow-water facies of the Sisquoc Formation in the Santa Maria basin (fig. 1) (Woodring and Bramlette, 1950:33) occur also in the Elsmere Canyon fauna. Since none of these species are lower Pliocene guides, Woodring and Bramlette (p. 102) considered the fossiliferous upper part of the Sisquoc Formation to be of middle Pliocene age. The lower part of the formation may be in part lower Pliocene, but those strata contain no megafossils.

Correlation of the Pancho Rico Formation in the Salinas Valley (fig. 1) with the Towsley Formation has been suggested by Durham and Addicott (1965:A17–A19). Of the 122 species of echinoids and molluscs in the Pancho Rico, 47 occur also in the Elsmere Canyon fauna.

The Tahana Sandstone Member and the Pomponio Sandy Mudstone Member of the Purisima Formation in the Santa Cruz Mountains (fig. 1) appear to be correlative with part of the Towsley Formation. Of 49 species of molluscs recorded in those members by Cummings, et al. (1962: pl. 24), 20 occur also in the Elsmere Canyon fauna. Cummings et al. (p. 211) suggested that the base of the Tahana Member is near the base of the Pliocene. Twelve of thirty-five definitely identified molluscan species in outcrops of the Tahana Member in Portola Valley (Addicott, 1969*b:* table 4) occur in the Elsmere Canyon fauna. Addicott (pp. 80–85) suggested an early Pliocene age for this assemblage.

The Pullen and Eel River formations of the Eel River basin in northern California (fig. 1) probably are correlative with the Towsley Formation, though the long geographic distance between these localities obscures the relationships, and the Eel River and Pullen faunas are small. Three of the six identified molluscan species in these formations (Faustman, 1964: fig. 7) occur also in the Elsmere Canyon fauna. Faustman (pp. 110–111) interpreted the Pullen and Eel River formations as being of early Pliocene age.

PICO FORMATION

Kew (1924:70–71) named the Pico Formation for exposures in Pico Canyon on the north slope of the Santa Susana Mountains (fig. 2). He intended to include all the rocks between the Modelo and Saugus formations. Rocks that he mapped as Modelo sandstone have since been placed in the Towsley Formation, which underlies the Pico Formation over much of the eastern Ventura basin. Winterer and Durham (1962:308–313) have described the lithology of the Pico Formation.

In upper Elsmere Canyon and south of lower Elsmere Canyon the basal part of the Pico Formation is a well-indurated conglomerate. This conglomerate interfingers with siltstone of the Towsley Formation in upper Elsmere Canyon, but elsewhere the contact may be unconformable (Winterer and Durham, 1962:308; Willis, 1952:36–37, 39). According to Winterer and Durham (1962:321), the Pico Formation in the eastern Ventura basin is entirely of Pliocene age. No fossils have been found in these rocks in the area of Elsmere and Grapevine canyons.

Pleistocene Series

Remnants of a post-Pico erosion surface are exposed on the Grapevine-Elsmere divide, and a small deposit of sedimentary breccia is exposed on one of the terraces cut into the lower slopes. In the southern half of the mapped area, where bituminous Pico conglomerate lies on soft, readily weathered siltstone, landslide deposits are common, even on gentle slopes.

STRUCTURE

Sedimentary rocks of the Towsley Formation in the area of Elsmere and upper Grapevine canyons dip westward at angles of 15° to 25° (fig. 4, in pocket). These rocks are separated from areas of more complex structure to the west and south by the Whitney Canyon and Grapevine Canyon faults.

The surface trace of the Whitney Canyon fault (Winterer and Durham, 1962: 337; my fig. 4) suggests that the fault is nearly vertical between Grapevine Canyon and the Placerita oil field, where it ends against the San Gabriel fault. Most of the movement on the Whitney Canyon fault apparently occurred before deposition of the Towsley Formation. The pre-Tertiary basement rocks are near the surface east of the fault but are overlain by a thick section of Eocene rocks west of the fault. Post-Eocene, pre-Pliocene movement on the fault produced a vertical separation of about 6,000 feet (Winterer and Durham, 1962:337; my fig. 8, in pocket, sec. B-B′). The east block moved up relative to the west, apparently forming part of a pre–San Gabriel range east of the mapped area (Miller, 1934:73–76). Vertical separation of the base of the Pliocene rocks, however, is only about 250 feet, in this instance with the west block relatively up, and the similarity of the rocks on opposite sides of the fault indicates a similarly small component of lateral separation. The small displacement of the Pliocene rocks may have occurred as local readjustment to stress produced at the time of movement on the Grapevine Canyon fault.

The temporal relationships of the faults in lower Elsmore Canyon are not clear. These faults are poorly exposed except at the western edge of the mapped area. The two northwest-trending reverse faults are thought to connect with the Legion fault, a major regional structural feature (Winterer and Durham, 1962: 336; my fig. 2).

The most recent uplift of this part of the San Gabriel Range apparently took place along the Grapevine Canyon fault. Northeast of this high-angle fault the base of the Towsley Formation is at the surface and dips steeply, but uniformly, to the southwest. The same contact southwest of the fault is overlain by a section of the Towsley Formation at least 1,500 feet thick (Winterer and Durham, 1962: 292). The thickness of the steeply dipping, folded, and faulted rocks southwest of the fault suggests vertical separation of 1,000 feet or more (fig. 8, in pocket, sec. D-D′).

Deformation along the Grapevine Canyon fault probably was contemporaneous with deformation that occurred during late Pliocene or early Pleistocene time in the San Gabriel and Sierra Madre fault zones (Hill, 1937:156: my fig. 2). At that time the Grapevine Canyon fault assumed the role of the Whitney Canyon fault as the major structural boundary of this part of the San Gabriel Mountains.

PALEOENVIRONMENTAL ANALYSIS

Methods, Materials, Assumptions

Paleoenvironmental interpretations in this study are based principally on the present-day distribution of fossil species. This method has two important limitations: fossil data may be biased in several ways, and environmental implications of species may change in time. These limitations are discussed below.

SAMPLE BIAS

The composition of a fossil collection usually is quite different from the composition of the living biota from which it was derived. The fossil sample may be biased in at least three different ways.

One source of bias lies in the preservation process. Organisms that do not have resistant skeletal elements are rarely preserved as fossils, and most marine benthic environments are dominated by soft-bodied organisms. Even shells of different composition or of different size may be selectively preserved or destroyed. Calcitic shells, for example, are much more resistant to leaching than are aragonitic shells. Consequently a fossil assemblage preserved in sedimentary rocks may contain only a small percentage of the species that originally lived in the sediments that formed those rocks.

A second source of error in sample interpretation lies in depositional processes. Skeletal elements of different sizes and shapes may be selectively transported before burial. Such partial assemblages, or even entire assemblages, may be transported and preserved in environments different from those in which the organisms lived. Such transported assemblages of course do not reflect environmental conditions in which they are preserved, and if their origin is not evident they may be misleading. Recognition of transported fossils is therefore essential for accurate interpretation of such local environmental conditions as depth of water, type of substrate, salinity, and so on.

Finally, the fossil sample may be biased by the collector. Nonobjective field methods in either collecting or observing can result in faunal lists that either do not include all species preserved in the rock or give incorrect relative abundance data. Such errors may be solely the fault of the paleontologist, though field conditions such as poor preservation, poor exposure, high degree of induration, and the like, may make his task very difficult.

In summary, the fossil sample on which the paleoecologist bases his reconstruction of the environment of deposition of a sedimentary rock may be very different from the biota that lived in the original sediments, and evaluation of the adequacy of the fossil sample is critical to such reconstruction.

Normally one cannot demonstrate the presence in a fossil fauna of organisms that have not been preserved, whether they had no hard parts or their hard parts were destroyed. Molluscs are by far the most abundant megafossils in the Elsmere Canyon fauna, yet a survey of the marine benthos off southern California shows that, in environments probably quite similar to those in which these fossils lived, polychaetes, crustaceans, and echinoderms are more abundant than molluscs (Allan Hancock Foundation, 1965:165). All fossil samples collected in this study are probably biased to some extent in this way.

Several criteria can be used to determine whether all or part of a fossil assemblage has been transported before burial and, in some instances, how far it has been transported (Johnson, 1960:1080). In this study it has been observed that in some beds the deeply burrowing bivalve *Panopea generosa* is preserved in life orientation, suggesting that in these beds at least the deep infauna is preserved in place. Very few fossil specimens are worn or abraded, further suggesting that there has been little transportation before burial. This interpretation is supported by the high percentage of bivalves that are preserved with the valves articulated (fig. 9, in pocket). It should be pointed out, however, that some species with strong ligaments may remain paired after the valves are broken, presumably by water turbulence, so articulation alone is not conclusive evidence of lack of transport. Finally, the ecological consistency of assemblages in different beds reflects their origin by burial in place without mixing from different environments. Dense accumulations of fossils have been thought to indicate concentration of shells by waves or currents, possibly from different environments, but the evidence described above strongly suggests that such accumulations of fossils in the Elsmere Canyon beds were preserved in place and that the fossil species can be used to reconstruct local environmental conditions. The origin of these beds is discussed below, as is specific evidence for transport or lack of transport of shells at different localities.

The fossil collections made during this study apparently have been biased to some extent by inadequate sampling of the preserved fossil assemblages. An at-

TABLE 2

COMPARISON OF SPECIES COMPOSITION AND ABUNDANCE IN DIFFERENT COLLECTIONS FROM TWO LOCALITIES

Locality L-2055				Locality 4782	
A	B	C		A	B
53	46	?	species 1	45	189
26	57	40	species 2	34	65
22	7	24	species 3	31	65
17	13	20	species 4	2	77
13	16	16	species 5	5	9
7	6	6	species 6	7	2
5	7	5	species 7	1	6
1	2	1	species 8	1	3
	5	6	species 9	3	1
4		4	species 10	1	1
4	3		species 11	1	1
1		4	species 12	15	
1		3	species 13	5	
1	1	3	species 14	5	
2	1		species 15	4	
2			species 16	3	
1	1	2	species 17	3	
1	1		species 18	2	
			species 19	1	
			species 20	1	
		3	species 21	1	
		3	species 22		3
		2	species 23		3
		1	species 24		1
		1	species 25		1
		1	species 26		1
		1	species 27		
		1	species 28		

NOTE: Collections A and B at locality L-2055 were made at different times during this study; collection C is from the UCLA invertebrate fossil collections. Collections A and B at locality 4782 were made at different times.

tempt was made to overcome this problem by making large collections and by collecting all identifiable specimens from bulk samples. These efforts were hampered by the high degree of induration of most beds, which made careful and thorough collecting a difficult and time-consuming process.

Collections were duplicated at two localities to test the reproducibility of the samples (table 2). Species composition and abundance in the collections from both localities suggest that the composition of samples taken in this study does not accurately reflect the composition of the fossil assemblages from which they were taken. These differences may also reflect, at least in part, the patchy distribution of many species within the bed.

Of the first obstacle to paleoenvironmental reconstruction—adequacy of the sample—it can be said that in this study all fossil collections are totally devoid of remains of soft-bodied organisms, there has been little or no transport of shells from one environment to another, and the fossil collections may not accurately reflect the composition of the fossil assemblages in the rocks.

UNIFORMITARIANISM

The second obstacle to paleoenvironmental reconstruction is the possibility that in some species the limits of tolerance to environmental variables may have changed through time. When the paleoecologist interprets past environments on the basis of the distribution of Recent species, he assumes that the environmental adaptations of those species have not changed in time. It is of course very difficult to check the validity of that assumption. If all the species in a fossil fauna can be found living together today, it may generally be assumed, at least for purposes of interpreting the environment in which that fauna lived, that none of those species have evolved ecologically. The presence in a fossil fauna of species that do not live together today would suggest that some of those species have evolved ecologically and that such evolution would render them less useful for purposes of environmental reconstruction. Specific patterns of distribution in space and time also may reveal ecological change in individual species. Recognition of ecological evolution is critical to paleoecological interpretation of the Elsmere Canyon fauna, and this problem is discussed at length below.

ENVIRONMENT

Environmental implications have been evaluated for fossil assemblages at each locality. Ecological data for all Recent species are included in the Systematic Paleontology section. The abundance of species and the inferred depth of water at each locality are shown in figure 9 (in pocket). The depth ranges of bathymetrically restricted species are shown graphically in figure 10, and water depths indicated by different assemblages are shown graphically in figure 11. See figure 4 (in pocket) for locations of fossil localities described in the following discussion.

DEPTH OF WATER

Lower unit.—The fauna of the basal conglomerate at UCLA localities 5768 and 5769 includes *Kelletia kelletii, Megasurcula carpenteriana,* and *Lucinoma annulata,* suggesting deposition in 10 to 25 fathoms of water (figs. 10, 11). Although the environment clearly was turbulent, the fauna does not appear to have been transported. The species present are typical of sand and boulder bottoms in shallow water off exposed shores.

The large clast size and the crude bedding of the conglomerate at UCLA localities 5785–5789 also suggest a turbulent environment. Water depths of 3 to 10 fathoms (fig. 11) are suggested by the presence of *Periploma planiuscula* and abundant *Amiantis callosa* (fig. 10), and by the absence of species that are restricted to water deeper than 10 fathoms. The shallow-water aspect is emphasized by the abundance of the exposed sandy-beach fauna, including *Tivela*

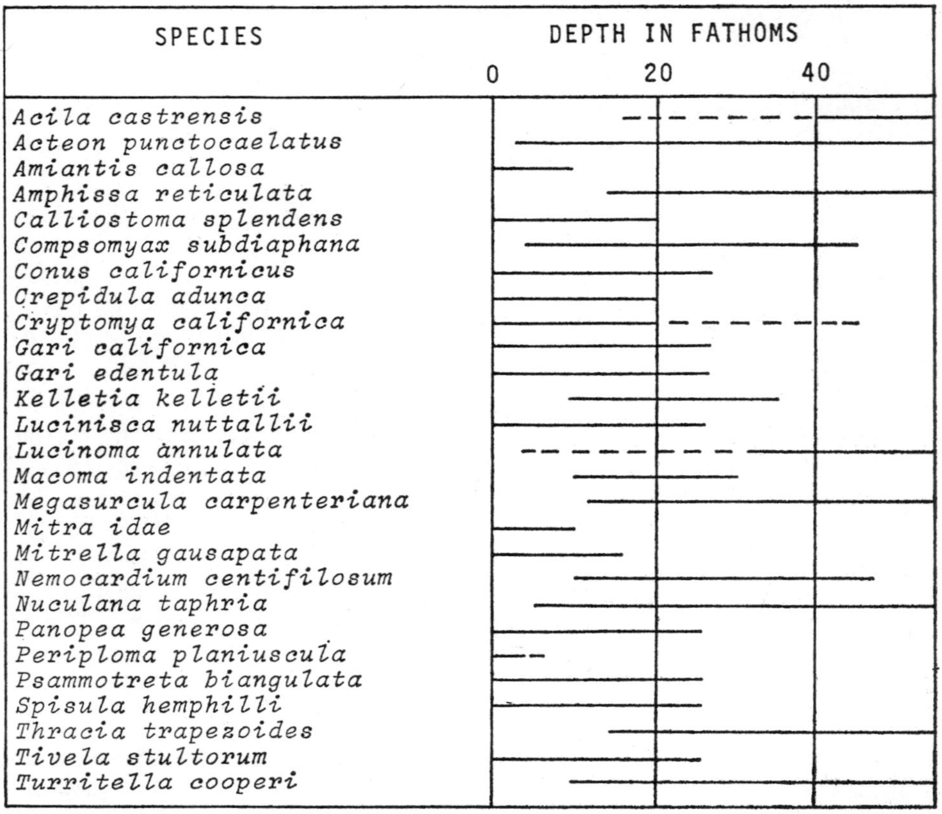

SPECIES

DEPTH IN FATHOMS

0 20 40

Acila castrensis
Acteon punctocaelatus
Amiantis callosa
Amphissa reticulata
Calliostoma splendens
Compsomyax subdiaphana
Conus californicus
Crepidula adunca
Cryptomya californica
Gari californica
Gari edentula
Kelletia kelletii
Lucinisca nuttallii
Lucinoma annulata
Macoma indentata
Megasurcula carpenteriana
Mitra idae
Mitrella gausapata
Nemocardium centifilosum
Nuculana taphria
Panopea generosa
Periploma planiuscula
Psammotreta biangulata
Spisula hemphilli
Thracia trapezoides
Tivela stultorum
Turritella cooperi

Fig. 10. Depth ranges of bathymetrically restricted species in the Elsmere Canyon fauna. Dashed lines indicate very low abundance. Sources of data are given in Systematic Paleontology section.

stultorum, Cryptomya californica, Macoma secta, Spisula hemphilli, and *Mactra* sp., and by the presence of several shallow-water rock dwellers, including species of *Astraea, Mitra, Littorina,* and *Acmaea.*

East of upper Elsmere Canyon the basal beds are less coarse-grained, and fossils from UCLA localities 5796–5799 suggest water depths of 12 to 20 fathoms (fig. 11). Basal beds elsewhere are mostly unfossiliferous, and the few small assemblages that do occur in these beds suggest water somewhat deeper than that in which the conglomerates were deposited. Apparently, as the sea transgressed on this shore, coarse sand and gravel were deposited locally in shallow water, and these deposits and the intervening rocky bottom were covered by more widespread sand and silt as the water deepened.

Fossil beds are distributed irregularly throughout the lower unit south of Elsmere Canyon. In general, these fossils suggest water depths of from 10 to 25 fathoms (fig. 11) and show little evidence of shell transport or disturbance. The fauna is that of a mud bottom; there are few species, and bivalves are predominant.

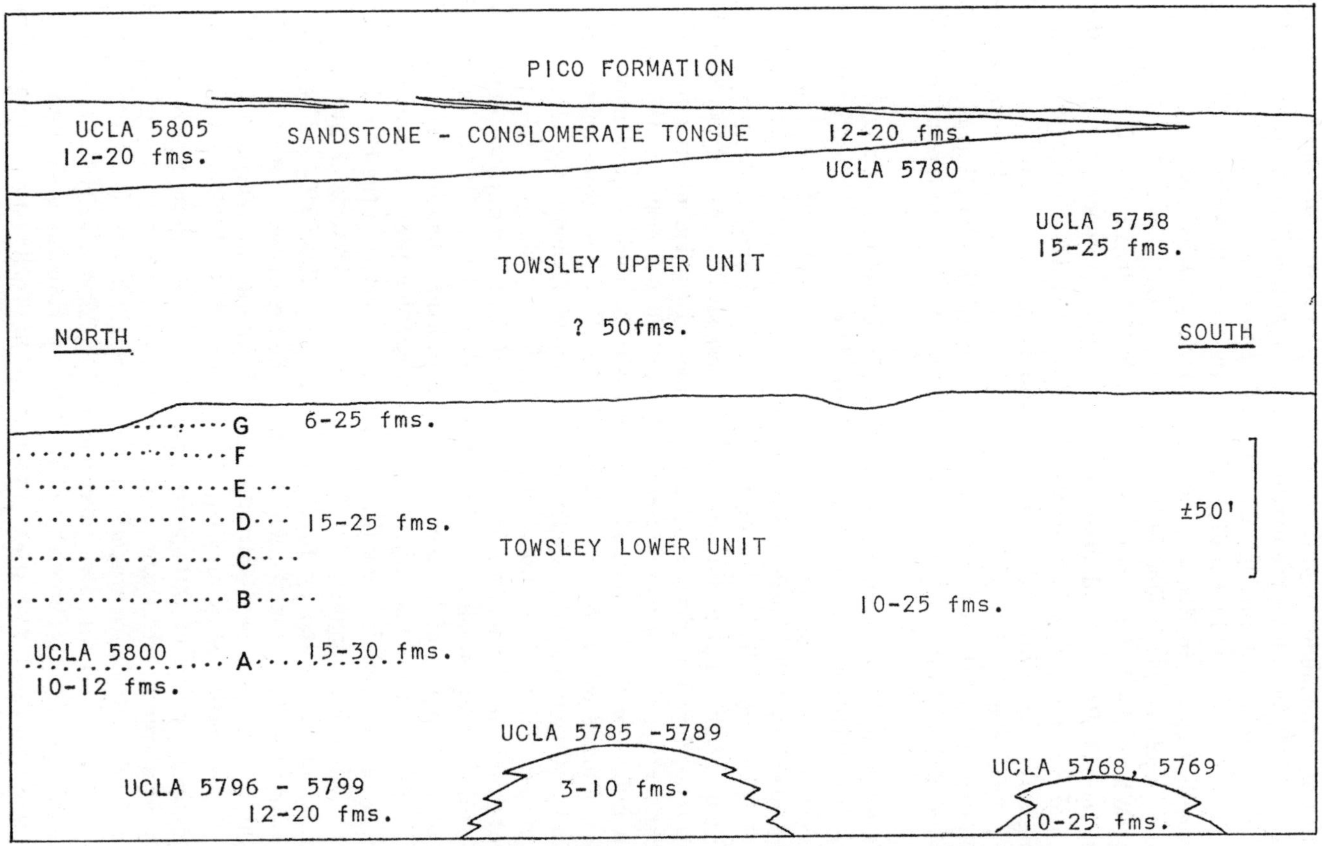

PICO FORMATION

UCLA 5805
12-20 fms.

SANDSTONE - CONGLOMERATE TONGUE

12-20 fms.
UCLA 5780

UCLA 5758
15-25 fms.

TOWSLEY UPPER UNIT

? 50fms.

NORTH

SOUTH

G 6-25 fms.

F

E

D 15-25 fms.

C

B

TOWSLEY LOWER UNIT

±50'

10-25 fms.

UCLA 5800
A 15-30 fms.
10-12 fms.

UCLA 5785 -5789
3-10 fms.

UCLA 5796 - 5799
12-20 fms.

UCLA 5768, 5769
10-25 fms.

Fig. 11. Diagrammatic illustration of water depths in which parts of the Towsley Formation in Elsmere Canyon were deposited. Stratigraphic thicknesses are drawn approximately to scale.

A series of seven highly fossiliferous beds in the lower unit can be traced for as much as 1,500 feet along the northwest wall of upper Elsmere Canyon. These are identified as beds A through G, from oldest to youngest (fig. 6, in pocket). Bed A is the most highly fossiliferous and the most laterally persistent. The abundance in this bed in Elsmere Canyon of *Acila castrensis, Macoma indentata,* and *Megasurcula carpenteriana* indicates water depths of 15 to 30 fathoms (figs. 10, 11). Where the bed extends eastward to UCLA locality 5800 *Acila* is not present, though *Macoma* and *Megasurcula* are. The presence here of *Amiantis callosa* and *Spisula hemphilli* suggests depths of 10 to 12 fathoms (figs. 10, 11), implying a difference in depth of 5 fathoms or more in a lateral distance of 400 feet, or a slope of more than 4° in the original bedding. If this slope was uniform, as the flat surface of the pre-Tertiary rocks suggests, the shoreline at the time of deposition of these beds was probably about 800 feet or less to the east of locality 5800.

The successively younger fossil-bearing beds B through F are not significantly different in faunal composition from bed A. *Acila castrensis* is locally abundant, as are *Megasurcula carpenteriana* and *Macoma indentata*. The depth of water during deposition of these beds was probably between 15 and 25 fathoms (fig. 11).

Bed G is the most distinctive faunal horizon, for *Turritella cooperi* is everywhere its most abundant molluscan species. These shells commonly show at least a slightly developed preferred orientation (see discussion of currents below). *Acila castrensis* and *Megasurcula carpenteriana* do not occur in this bed, and *Macoma indentata* is rare. The absence or rarity of these species may indicate slightly shallower water than that in which the older beds were deposited. The fauna in bed G suggests water depths of from 6 to 25 fathoms (fig. 11).

Upper unit.—Environmental conditions during deposition of the upper unit apparently were similar to those in which most of the lower unit was deposited. A diverse and abundant fossil fauna in a thin sandstone bed at UCLA locality 5758 indicates water depths of between 15 and 25 fathoms (fig. 11). UCLA locality L-2056 is probably in this unit, but its stratigraphic position is uncertain. It appears to be very high in the unit, close to the contact with the Pico Formation. The depth of water suggested by this fauna is from 12 to 20 fathoms; similar depths also are suggested by fossils at UCLA localities 5780 and 5805 in the sandstone-conglomerate tongue, which is near the top of the upper unit (fig. 11). A foraminiferal fauna has been reported by Winterer and Durham (1962:307) from a single locality in the upper unit just west of the Whitney Canyon fault. Living representatives of the species reported live at depths greater than 100 fathoms. Winterer and Durham were uncertain as to the accuracy of this interpretation, and the evidence presented here apparently is opposed to it. Megafaunal evidence is lacking in this area, however, and it is possible that the water was somewhat deeper during deposition of the lower part of the upper unit.

Thus the fossil fauna reflects no large changes in depth of water during deposition of sediments of the Towsley Formation in this area. Sedimentation was initiated in water less than 10 fathoms deep, but apparently the depth increased to 25 or 30 fathoms before sea level became stabilized. The water may have become somewhat more shallow during deposition, and the upper part of the lower

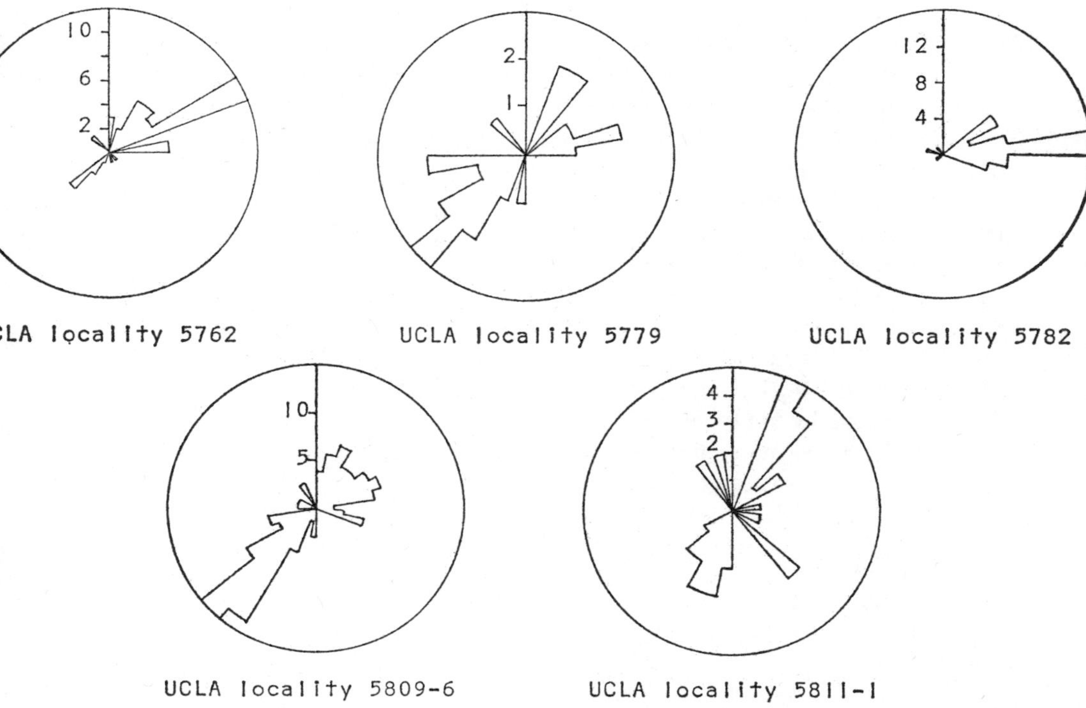

Fig. 12. Orientation of *Turritella cooperi* shells at five localities. The direction of the apex is plotted.

unit was deposited in water probably less than 20 fathoms deep. Depth of water during deposition of the upper unit probably was between 12 and 25 fathoms.

CURRENTS

Oriented shells of *Turritella cooperi* at UCLA localities 5762, 5779, 5782, 5809-6, and 5811-1 apparently reflect the action of weak to moderately strong currents. The majority of the apexes of shells at these localities point to the east-northeast, southwest, and north-northeast (fig. 12). Probably the shells are aligned parallel to current direction, but the sense of current flow relative to shell orientation is not clear (Potter and Pettijohn, 1963:37–39). The orientation of these shells suggests a roughly northeast–southwest direction of current flow, approximately parallel to the inferred early Pliocene shoreline. Perhaps they were aligned by longshore currents flowing either northeast or southwest. Cut-and-fill structures, current bedding, current lineations, and ripple marks in the Towsley Formation along the north slope of the Santa Susana Mountains show a remarkably constant direction of current flow from the east-northeast (Winterer and Durham, 1962: 330). These currents were in deep water, however, and probably were not related to nearshore currents.

ORIGIN OF FOSSIL BEDS

Fossils are densely concentrated in beds A and G and at least locally in the other beds in upper Elsmere Canyon. Shell material commonly makes up as much as

25 percent of these rocks. A high percentage of the bivalves are paired. (Data are presented in fig. 9, in pocket.) The shells, usually unbroken and unworn, show no sign of significant postmortem transportation, and the ecological coherence of the fauna at any given locality also indicates deposition in place.

In Recent shell concentrations in Tomales Bay, California, many of the abundant shells have articulated valves and there is little shell debris (Johnson, 1965: 83). Among the substrates sampled, the mud bottoms have the highest concentrations of shells, but they contain the lowest densities of living animals. Currents are weak, and the rate of deposition of sediment is low. These Recent shell concentrations apparently are produced by a small living population combined with a low rate of deposition.

The fossil beds in Elsmere Canyon probably had a similar origin. Alternating intervals of deposition and nondeposition could account for the series of shell beds separated by zones in which shells are present but not concentrated. Thus faunal abundance may have been relatively constant during deposition of the sediments forming these rocks.

The absence of large numbers of fossils in the lower unit south of Elsmere Canyon and in the upper unit may be due to the finer grain size of the sediments. Most of the fossils in these rocks occur in thin beds of sandstone and conglomerate. Muddy substrates generally support a less abundant and less varied molluscan fauna than even very fine sand (for example, see Smith and Gordon, 1948: 154–155).

PALEOCLIMATIC ANALYSIS

Methods, Assumptions, and Problems of Paleoclimatic Interpretation

The usual approach to interpretation of paleoclimate using fossil distribution is based on changes in faunal composition with latitude. The geographic distribution of nearly all marine invertebrate species is more or less limited along north–south coastlines. Consequently the total invertebrate fauna is characterized by latitudinal change in species composition along such coasts. This faunal change parallels the latitudinal gradient in temperature. The total fauna at any point is unique to a short stretch of coast north and south of that point, and the temperature regime along the same stretch is unique to that particular north–south coastline. Thus there is a direct correlation between faunal composition and thermal conditions along the entire coast.

A complete faunal list from any point along the coast tends to define the latitude, and therefore the thermal regime, where that fauna lives. The overlapping geographic ranges of component species of the fauna usually can be so used to locate the source of the fauna very accurately. This method is generally used in paleoclimatic interpretation. The thermal regime under which a fossil fauna lived is presumed to be the same as the thermal regime at the point or area of overlap of the present-day geographic ranges of all the species in the fauna. A fossil fauna in a place different from the area of overlap of the present-day geographic ranges of the species in that fauna is thought to indicate a corresponding climatic change. Specifically it is thought to indicate that the thermal regime at the time when and the place where that fauna lived was the same as the thermal regime today at the place of overlap of the present-day geographic ranges of those species.

This approach to paleoclimatic interpretation is based on two assumptions. The first is that the geographic distribution of marine animals is limited by temperature. This assumption has been stated and endorsed by a large number of biologists and paleontologists, including Hutchins (1947:325), Bullock (1955: 312–313), Gunter (1957: 159, 172), Valentine (1961:315), and Kinne (1963:317). The second assumption is that changes through time in the geographic distribution of invertebrate species reflect changes in environment and not physiological changes in animals. This assumption also has long been regarded by paleontologists as reasonable, at least for young fossils. For example, Valentine (1955:466), speaking of Pleistocene fossils, has stated that "the possibility that evolutionary changes in the physiology and therefore ecologic requirements of species have enabled them to alter their environments also has been considered. . . . This is certainly possible for a few species, but such changes in large numbers of forms appear improbable."

The most outstanding, and to paleoecologists the most frustrating, fact conflicting with this assumption is the presence in many fossil faunas, and in most California Pleistocene and Pliocene fossil faunas, of a few thermally anomalous species, species whose present-day geographic ranges do not overlap with those of the majority of the species with which they are associated. Many of these faunas include both species that today live north and others that live south of the common range of associated species. A number of explanations (which are discussed below) have been proposed for such associations of thermally anomalous species, and several of these may be correct, at least in part.

I feel, however, that a satisfactory evaluation of this problem requires an understanding of the complex relationships between environmental temperature and the distribution of marine poikilotherms (animals whose body temperature is not much different from their surroundings) and that it will be instructive to review current knowledge of these relationships. This review also includes information possibly bearing on the second assumption, that the temperature requirements of species have remained unchanged through time.

Water Temperature and Distribution of Marine Poikilotherms

Temperature and Survival

In a latent state protoplasm can survive temperatures as low as –270°C and as high as 150°C (Kinne, 1963:305). In an active state, however, few marine poikilotherms are able to survive in water warmer than about 35°C or colder than a few degrees below 0°C, though some intertidal species can withstand temperatures as high as 45°C (Gunter, 1957:163; Kinne, 1963:305–306; Raymont, 1963:8). Most species, moreover, are restricted to temperature ranges narrower than these extremes (Kinne, 1963:305).

Mortality of marine poikilotherms at both high and low temperatures in nature has been documented by Brongersma-Sanders (1957), Gunter (1957), Kinne (1963:305–306; 1970:410–413), and others. The causes of death from extreme temperatures are complex and not well understood. Kinne (1963:308) suggested that death at extremely high temperatures may be owing to reduction in oxygen supply, failure in process integration, desiccation, inactivation of enzymes, change in state of lipids, increase in protoplasmic viscosity, or increased permeability of cell membranes. At higher temperatures protein denaturation and liberation of toxic substances may occur. Kinne suggested that death at extremely low temperatures may be caused by insufficient rates of energy liberation, changes in water and mineral balance and in colloidal relations with water, dehydration of cells as a result of extracellular freezing and increase in osmoconcentration, and liquefaction of cortical protoplasm and gelation of the interior. Death may also occur, however, as a result of interference with rates of and integration between nervous and metabolic processes at less extreme high or low temperatures before any protoplasm deterioration occurs (Kinne, 1963:308; 1970:330; Raymont, 1963:18). Dickie (1959:73) has suggested that death at high temperatures may be caused by reduced mobility, resulting in starvation, or by increased predation. The restriction of most poikilotherms to rather narrow temperature ranges further suggests that thermal death limits are determined not by loss of viability of protoplasm at extreme temperatures but by loss of capacity to carry out metabolic processes at less extreme temperatures. Thus the effects of extreme temperature on survival may be quite varied in different species, but in general extreme temperatures are lethal in one way or another, and survival temperature is at least a potentially limiting factor in distribution of marine poikilotherms.

Species whose geographic distribution is limited by survival temperature have simple temperature-distribution relationships. Northern range end points are set by winter temperatures; water north of range end points is too cold for survival in winter. Southern range end points are set by summer temperatures; water south of range end points is too warm for survival in summer. (The compass directions used in this discussion apply in the northern hemisphere; "north" and "south" would be interchanged in the southern hemisphere.) If all poikilotherms were limited by survival temperatures, the ranges of fossil species would define the past positions of winter isotherms in the north and summer isotherms in the south.

TEMPERATURE AND LIFE PROCESSES

The critical importance of temperature to metabolic and other processes suggests that the distribution of some marine poikilotherms may be restricted by temperature requirements in aspects of life other than survival. Kinne (1963:317–318) has proposed that temperature may limit the distribution of species by effects on rates of growth and metabolism, on intra- and interspecific relationships, and on various aspects of reproduction and early development of offspring.

Appellöf (1912:556–557) and Orton (1920:342–345, 356–357) were the first to suggest that the distribution of some marine organisms may be limited by temperature requirements for reproduction. In more recent studies, Hutchins (1947:325), Bullock (1955:312–313), Hedgpeth (1957:375–376), Valentine (1961:

316), and Kinne (1963:305, 317–318; 1970:410) have suggested that such limitations may be imposed by temperature requirements for spawning, larval growth, larval metamorphosis, or other phases of the reproductive process. Reproduction in some organisms continues as long as the temperature is above or below some critical level. High-latitude species may breed only during the winter in the southern parts of their ranges and all year in the northern parts, and low-latitude species may breed only during the summer in the northern parts of their ranges and all year in the southern parts. Reproduction in some other organisms occurs only between two critical temperatures. Such species breed only in winter in the southern parts of their ranges and only in summer in the northern parts.

Species whose geographic distribution is limited by temperatures required for reproduction have northern range end points set by summer temperatures and southern range end points set by winter temperatures. Water north of northern range end points is not warm enough for reproduction in summer; water south of southern range end points is not cold enough for reproduction in winter. If all poikilotherms were limited by temperatures required for reproduction, the ranges of fossil species would define the past positions of summer isotherms in the north and winter isotherms in the south.

There also is evidence that aspects of the annual thermal regime other than simple minimum and maximum temperatures may limit the distribution of some organisms through effects on reproduction. Hedgpeth (1957:375–376) and Crisp and Southward (1958:188) suggested that temperatures must be above or below a given value for a critical length of time for reproduction of some species to be successful. For example, the southern range end point of *Balanus balanoides* in Europe may be related to the necessity for a sufficiently long period of low temperature to begin breeding (Crisp, 1957:1139).

As further noted by Kinne (1963:304; 1970:330), the biological effects of a given temperature on any given species may differ in different populations, at different ages, at different life stages, and in different sexes. In evaluating the effects of temperature on distribution one must take into account past and present patterns of variation; distinctions among constant and fluctuating temperatures, temperature gradients, ranges, and averages; frequency and intensity of changes; duration of given patterns; and total summation in time. Bullock (1955:333) cautioned that "without further work we cannot assume that the monthly average, daily average or any given measure short of the complete curve of temperature against time, is biologically appropriate, since we do not know how the organism weights equal departures from the average in the two directions or for equal lengths of time."

INTERPRETATION OF TEMPERATURE-DISTRIBUTION RELATIONSHIPS

It is clear, then, that the relationships between the environmental thermal regime and the geographic distribution of marine poikilotherms are very complex, and the distribution of the different species present at a given latitude may be limited by different kinds of temperature requirements at different seasons of the year and in different parts of their life histories. The northern and southern range end points of a given species may be determined by any of various temperature effects on survival, growth, reproduction, and so on. Ideally the temperature-distribution relationships of each species in a fossil assemblage should be known to best interpret the thermal significance of the assemblage.

INDEPENDENCE OF TEMPERATURE AND DISTRIBUTION

General statement.—There is some evidence suggesting that the geographic distribution of many marine poikilotherms is not directly limited by temperature. Although most such organisms are restricted in geographic range, all parts of the sea have abundant life, and closely related species commonly thrive under quite different thermal conditions (Thorson, 1957:505–521). This thermal separation of closely related species suggests that geographic ranges may not be limited by genotypic differences between species. That is, they may not be limited by genetically controlled temperature tolerance limits. Independence of geographic distribution from temperature control also has been suggested by Crisp and Southward (1958:187–188), who found that geographic limits of intertidal organisms in the English Channel do not correspond with isotherms at any season. Further, Raymont (1963:306) has pointed out that many planktonic animals seem able to tolerate wider ranges of temperature than those to which they are subjected within their normal geographic ranges.

This idea has recently been supported by Kinne (1970:502–503): "Distributions of invertebrates in the sea are the result of a multitude of environmental and organismic properties, interrelated in various ways and all subject to changes in time. Consequently, attempts to explain distributions must consider a variety of historical as well as present-day aspects. Even if such considerations are restricted to the present situation, the causes underlying distributions in the sea are presumably so complex that temperature can be expected to represent the primary controlling factor only if (i) habitats are compared with widely different thermal regimes, (ii) attention is focused on areas with extreme or significantly changing temperatures, or (iii) organisms are considered which have rather specific temperature requirements.

"Even a geographically perfect match of organismic abundance and temperature conditions can only provide circumstantial evidence for a causal relationship between distribution and temperature."

Acclimation.—Bullock (1955:317, 321, 336) has suggested that many species of poikilotherms exhibit in their metabolism or activity some degree of independence of their temperature, and that adaptation of metabolism and other rate functions to latitude and to seasons as compensation for, or acclimation to, temperature is widespread in nature. Cold-adapted individuals have higher metabolic rates than warm-adapted individuals at the same temperature. This relationship holds for individuals of the same species living in cold and warm climates and in different depths of water and also for seasonal temperature changes in nature and experimental temperature changes in the laboratory. Such acclimation is well developed in many poikilotherms, whereas in others it apparently does not occur.

Many species of marine poikilotherms are able to shift their tolerance levels in response to temperatures normally encountered at a given place and time (Segal,

1961:242). Lethal temperatures for most species are directly related to the maximum and minimum temperatures of the environment. Lethal minimum temperature is lower in winter than in summer for the same species, and lethal maximum temperature is higher in summer than in winter. Metabolic rates, and presumably also lethal temperatures, are different at different latitudes and at different depths of water for the same species (Bullock, 1955:336; Segal, 1956:146). Thus most species adapt to spatial and temporal changes in the thermal regime within their geographic range and are not limited by specific minimum and maximum temperatures or other thermal variables.

Adaptation to latitudinal temperature gradients also is reflected in differences in reproductive physiology between tropical and polar species. Arctic and Antarctic invertebrates more commonly have large eggs or are viviparous and more commonly brood their young; and planktonic larval stages, which are more susceptible to temperature extremes than adult stages, occur in only 5 percent of Arctic species compared with 65 percent of temperate species (Giese, 1959:555–559). Thus many polar species escape the limiting effects of extreme temperatures.

The limiting effects of temperature on geographic distribution seem to have been overcome in a variety of ways by many species of marine poikilotherms. The geographic ranges of many of these species may then be limited by factors other than temperature effects on survival, reproduction, and other life processes.

Physiological species.—Appellöf (1912:555) first suggested that some widely distributed species actually may consist of several physiologically distinct races, each morphologically indistinguishable from the others but adapted to a different thermal regime. Korringa (1957) applied this concept to ***Ostrea edulis*** in Europe, recognizing three physiological races that begin to release their larvae at environmental temperatures of 17°, 20°, and 25°C, respectively. Stauber (1950) and Loosanoff and Nomejko (1951) described three physiological races of ***Crassostrea virginica*** which spawn at different latitudes on the east coast of North America at minimum temperatures of 16.4°, 20°, and 25°C. Stauber (1950) also described two physiological races of the oyster drill ***Urosalpinx cinerea,*** a predator of ***Crassostrea virginica;*** these races spawn at temperatures of 10° and 20°C. These conclusions have been questioned by Galtsoff (1961:281–282), who found in laboratory studies that all individuals of ***C. virginica*** at a given locality do not spawn at the same temperature but that they require stimulation by sperm. Prosser (1955:241–244) suggested that physiological races exist in ***Paramecium aurelia, Daphnia, Drosophila, Rana,*** and various tunicate and echinoderm species. Axelrod (1941) has employed the same concept in suggesting that Tertiary plant distribution patterns reflect the existence of separate ecological species within given plant species populations (see below).

There seems to be substantial evidence, then, that some poikilotherm species consist of distinct physiological races that are adapted to different thermal conditions. Whether or not these races are genetically distinct, they must be very closely related and they are acclimated to different thermal conditions, further suggesting that distribution of such species may be independent of temperature.

Time, temperature, and reproduction.—The three physiological races of ***Crassostrea virginica,*** which Stauber (1950:112–113) found spawning at different temperatures all begin to spawn at about the same date. The same is true for *Urosalpinx cinerea.* Giese (1959:556) found that while many tropical species are seasonal, others breed throughout the year. There is also evidence that different species breed at different times of the year in temperate seas with the result that larval forms are present throughout the year (Pyefinch, 1948:465–493; MacGinitie, 1955:110–111; Giese, 1959:553–554).

Thus seasonal reproduction of some species may be a mechanism to reduce competition for food or other necessary resources. Though such species probably are adapted to environmental temperature regimes (see below), their reproductive processes and geographic distribution probably were not limited initially by temperature. Further, the seasonal abundance of larvae in many seas corresponds to the abundance of food, and some species can breed anytime there is sufficient food for larvae (Thorson, 1946; Giese, 1959:553, 556; Raymont, 1963:307). Such species clearly are not limited by temperature requirements for reproduction.

Changes in temperature-distribution relationships in the fossil record.—Temporal changes in the relationships between temperature and geographic distribution of a number of marine invertebrate taxa are documented in the fossil record. Several examples are described below in the discussion of Elsmere Canyon paleoclimate. Such changes suggest either that the distribution of some invertebrate species is limited by factors other than temperature, or that temperature requirements have changed through time.

Adaptation of species to accustomed temperature.—There is, then, evidence suggesting that the distribution of at least some marine poikilotherm species is not limited directly by temperature. Yet most species are restricted within rather narrow latitudinal ranges, and most species are adversely affected by temperatures outside the range to which they are normally subjected. Clearly there is a close relationship between distribution and environmental temperature.

According to Raymont (1963:10), the "species peculiar to a marine area have become adapted so that their physiological processes work most efficiently at the temperature levels normally experienced in the particular marine environment." Assuming that the geographic range of a species may be limited by factors other than temperature, the species naturally would become adapted to the thermal regime within that geographic range. Temperatures outside the normal range of that regime would have adverse effects on that species, especially if abrupt temperature changes occur.

Random yearly fluctuations in thermal regime may cause mass mortality, suggesting direct temperature control of distribution, but such mortality may occur anywhere in a species' range, not just at range end points. Mortality may be a result of sudden extreme temperature change rather than of temperatures beyond the limits of tolerance for the species (Kinne, 1963:305–306). Individuals of a species whose range is limited by factors other than temperature may be killed by sudden extreme temperature change, though the species' range is not directly limited by temperature.

The close relationships between environmental temperatures and reproductive cycles of many marine poikilotherms also may reflect adaptation to accustomed conditions. Further, such relationships may serve in some species as a mechanism for maintaining the seasonal patterns of reproduction. As I suggested above, breeding periods of different species may be distributed during the year to utilize available food most efficiently or to maintain the stability of predator-prey relationships. The temperature stimulus to reproduction may have evolved in some species as a mechanism for maintaining this timing between the reproductive seasons of different species.

Biological control of distribution.—Crisp and Southward (1958:187) suggested that distribution may be limited by competitive relationships between species. The role of competition and other biological interactions between species in limiting their distribution is little understood. The distribution of some species may be limited by concentrations of vitamins, antibiotics, or other organic metabolic by-products in trace quantities in the water (Lucas, 1947; 1955; 1961; Raymont, 1963:306–307). These substances, called external metabolites, may enhance or inhibit the viability of larvae or adults of certain species. Very little is now known about the effects of these substances on the distribution of organisms.

SUMMARY

The distribution of individual species of marine poikilotherms is limited in geographic directions in which there exist temperature gradients. Geographic distribution of some species may be limited directly by the effects of temperature on survival, on growth, on reproduction, or on other life processes. Such limitations, however, may be imposed not only by minimum or maximum seasonal temperatures but also by temperature changes, durations, or other aspects of the seasonal thermal regime. Thus temperature-distribution relationships may be very complex in species that are limited directly by temperature.

Some species, however, may not be limited by temperature. They may be limited, for example, by biological interactions with other species. Most such species, however, must become adapted to the thermal regime within their geographic range, so the distribution of most species correlates closely with temperature patterns. But the correlation may be tenuous, and many of these species presumably can alter temperature tolerance ranges in relatively short periods of time, while some species may be almost entirely independent of temperature control.

Paleoclimatic interpretations based on the present-day distribution of fossil poikilotherms must take into account the complexity of the relationships between the distributions of these organisms and environmental thermal regimes and also the relative or total independence of some species from temperature control, especially over intervals of geologic time.

Biogeographic Elements in the Elsmere Canyon Fauna

This fauna is a critical one in the reconstruction of Tertiary marine paleoclimate on the west coast of North America because it is the only large shallow-water fauna of early Pliocene age in Southern California. Three different biogeographic elements are present in the fauna. These elements are compared with faunas of present-day molluscan provinces as defined by Valentine (1966). A southern element includes taxa living today in the Surian province (Cape San Lucas to Cedros Island) and in the Panamic province (south of Cape San Lucas). The dominant element in the Elsmere Canyon fauna represents the present-day Californian province (Cedros Island to Point Conception). A northern element includes taxa living today in the Oregonian province (Point Conception to 56° north latitude) and in the Aleutian province (56° north latitude to Prince William Sound). There is no stratigraphic change in relative importance of these elements in the Elsmere Canyon section.

CALIFORNIAN FAUNAL ELEMENT

The geographic ranges of all specifically identified Recent molluscs with ranges that end near or do not include the latitude of Elsmere Canyon are shown graphically in figure 13; the geographic ranges of all the positively identified Recent species in the fauna are shown in figure 14 (in pocket). Twelve identified Recent species (fig. 13) have southern shallow-water range end points within 2 degrees of the latitude of Elsmere Canyon; four other southern range end points are within 3 degrees and three more within 4 degrees of Elsmere Canyon. Thirteen species have Recent northern range end points within 1 degree of Elsmere Canyon and three others within 3 degrees. Of the sixty-seven identified Recent species in the fauna all but five (which are discussed below) have shallow-water ranges coincident between 33° and 34° north latitude. The position of this zone of overlapping ranges relative to the latitude of Elsmere Canyon (34° 20′) suggests early Pliocene climate slightly warmer than at that latitude today, possibly similar to present-day climate near the latitude of Newport Beach (33° 30′).

Of the twenty-one questionably identified species, *Bankia setacea, Acmaea digitalis,* and *Bittium rugatum* have southern range end points in this region; the latter species and *Clavus pallidus* have northern range end points here. The ranges of all but two (which are discussed below) of the remaining questionably identified taxa include the latitude of Elsmere Canyon.

NORTHERN FAUNAL ELEMENT

This small element of the Elsmere Canyon fauna includes four living and two extinct species. The Oregonian province is represented by five species. *Natica clausa* lives in shallow water only north of 37° north latitude but lives off southern California in deep water (fig. 13). *Mya truncata, Neptunea lyrata,* and *Fusitriton oregonensis* live in shallow water only north of Puget Sound (48° north latitude); the latter two species range into deep water off southern California. *Patinopecten lohri* is extinct; living species of this genus live only north of Point Reyes at 38° north latitude (Burch, 1944–1946: no. 35:8). *Crepidula grandis* Middendorff of the Recent Aleutian province may be a descendant of the extinct *C. princeps.*

The four Recent species are rare in the Elsmere Canyon fauna (fig. 9, in pocket), I collected six specimens of the *Fusitriton* and one each of the *Mya* and *Neptunea;* a single specimen of *Natica clausa* was reported by Grant and Gale (1931:797). *Crepidula princeps,* however, is common, and *Patinopecten lohri* is abundant.

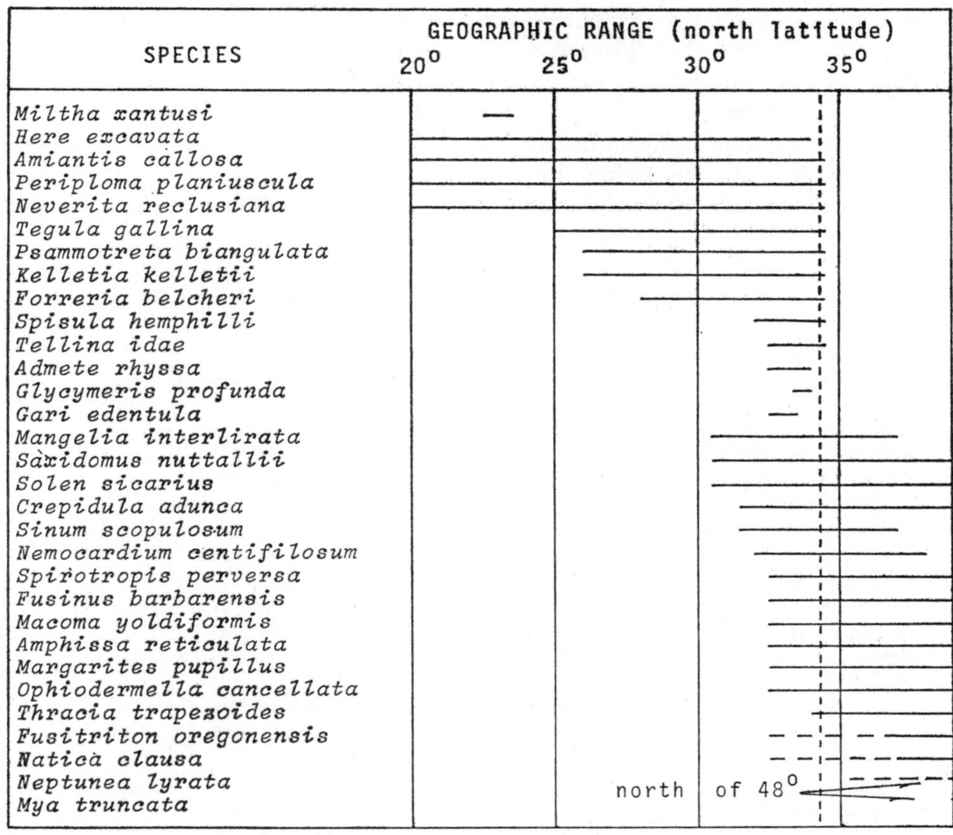

Fig. 13. Shallow-water geographic ranges of species that have present-day range end points near the latitude of Elsmere Canyon. Dashed lines indicate extensions of ranges in deep water. Vertical dashed line is latitude of Elsmere Canyon. Sources of data are given in Systematic Paleontology section.

SOUTHERN FAUNAL ELEMENT

Listed in table 3 are fourteen species (in thirteen genera) that are treated tentatively as representative of the Recent Surian and Panamic molluscan provinces. Three of these taxa are questionably referred to living species and one is known to be living; all others on the list are extinct. Except for *Astele* aff. *A. rema, Cantharus* sp., and *Cymatium elsmerense,* these species are moderately abundant in Elsmere Canyon (fig. 9, in pocket). The greater diversity and abundance of the southern element relative to the northern suggest that the southern extralimital taxa are of greater paleoclimatic significance.

Extralimital and Thermally Anomalous Fossil Taxa

Thermally anomalous taxa are those that live today under thermal regimes warmer or colder than those suggested by the majority of the taxa with which they are associated in fossil faunas (Valentine, 1955:463–465). The term "extralimital" has been used by Valentine and Meade (1961:1) for taxa that occur in fossil faunas beyond their present-day geographic ranges.

Although the dominant element in the Elsmere Canyon fauna suggests a marine thermal regime slightly warmer than that at the same latitude today, there are two smaller extralimital and thermally anomalous elements in the fauna. Six species, four of them living, apparently represent the present-day Oregonian and Aleutian molluscan provinces and suggest a thermal regime colder than that existing today at this latitude. Twelve to fourteen species, at least one of them living, apparently represent the present-day Surian and Panamic molluscan provinces and suggest a thermal regime warmer than that existing today at this latitude.

TABLE 3

Elsmere Canyon Molluscan Taxa That Have Affinities with the Fauna of the Surian and Panamic Provinces

Taxon	Present-day geographic range of genus*
Anadara trilineata	Newport Bay to Peru
Arca terminumbonis	Scammon's Lagoon to Peru
Chione (Anomalocardia) fernandoensis	Magdalena Bay to Peru
Dosinia jacalitosana	Scammon's Lagoon to Peru
Miltha xantusi	Cape San Lucas south (*M. xantusi*)
Lyropecten estrellanus	Scammon's Lagoon to Ecuador
Cymatium elsmerense	San Ignacio Lagoon to Peru
Astele aff. *A. rema*	Tres Marías Islands to Mazatlan
Cantharus sp.	Magdalena Bay to Peru
Calicantharus humerosus	Extinct (related to *Cantharus*)
C. fortis angulatus	Extinct
Ficus (Trophosycon) ocoyana	Cape San Lucas to Ecuador
Bulla cf. *B. punctulata*	Magdalena Bay to Peru (*B. punctulata*)
Olivella aff. *O. gracilis*	Guaymas to Peru (*O. gracilis*)

*Geographic ranges from Keen (1958).

Such mixing of biogeographic elements in fossil biotas is common in west coast Tertiary and Quaternary deposits (see below) and also occurs in Europe, Japan, and elsewhere (Axelrod, 1967*b*:5–7). It was first recognized in California by Carpenter (1866), who commented on the mixing of warm-water and cool-water species in Pleistocene faunas at Santa Barbara. Carpenter gave no explanation for this mixing of biogeographic elements, but explanations have been sought since by a number of paleontologists, including Crickmay (1929), Axelrod (1941, 1967*b*), Woodring, Bramlette, and Kew (1946), Woodring (1951), Valentine (1955, 1961), Emerson (1956), and Valentine and Meade (1961). These explanations may be summarized as follows: (1) fossils from different environments may be mixed; (2) some species may have changed their thermal requirements or adaptations through evolution; (3) distribution of some species may be independent of temperature; (4) changes in faunal distribution may reflect environmental (thermal) changes in time.

Possible Explanations for Thermally Anomalous Fossil Faunas

MIXING OF FOSSILS FROM DIFFERENT ENVIRONMENTS

Crickmay (1929:621–632) attributed the mixing of cold-water with warm-water fossil species in Pleistocene deposits on Deadman's Island to reworking of fossils from older formations. Such a process may account for thermal anomalies in isolated cases in which successive formations of nearly the same age were deposited under different climatic conditions, a situation that does occur in the California Pleistocene section. Reworked fossils usually can be recognized, however, particularly if there is a substantial difference in age between the formations involved, and clearly most of the thermally anomalous faunas of the California Tertiary, including the Elsmere Canyon fauna, cannot be explained in this way.

Woodring, Bramlette, and Kew (1946:101–102) suggested that some mixing of northern and southern fossil species may be a result of mixing of depositional facies, especially shallow-water and deepwater facies. This process probably does not account for many of the instances of thermally anomalous fossil faunas, and one should be able to detect evidence in a fossil assemblage of such an origin. There is no evidence for such mixing in the Elsmere Canyon fauna.

ECOLOGICAL CHANGES IN TAXA

Axelrod (1941:546) suggested three possible explanations for mixtures of plants from widely different habitats in Tertiary fossil floras of western North America. Two of these explanations are predicated on environmental changes, while the third, based on the concept of the physiological species first proposed by Appellöf (1912:555), supposes the existence within a given species of physiological races or ecospecies living in regions of different environmental conditions. Late Tertiary plant species closely related to modern endemics that have rather narrow ecologic limits apparently lived under diverse climates and with widely different types of vegetation. Extinction of some of these morphologically indistinguishable ecospecies might have resulted in a "species" having a Recent geographic range quite different from its range as a fossil. Though fossil plants may be indistinguishable from living species, some may have been environmentally wide-ranging ecospecies rather than ecologically restricted endemics. Axelrod (1941:548) suggested that such ecospecies may be a result of polyploidy, a phenomenon common to plants but not to animals. Nevertheless, the existence of sibling species in many different kinds of animals suggests that such a concept may also be applicable to animal species. This concept essentially proposes a possible mechanism for changing the environmental tolerance limits of a species, or of what would be called a species by a paleontologist. Changes in distribution would reflect evolutionary changes in the species rather than changes in environment.

Woodring (1951:482–483) attributed the mixing of thermal elements in late Pliocene and Pleistocene rocks of California to two factors, the "rise and extinction of local subspecies, or species, differentiated by physiological characters," and "subsequent evolution in physiological characters that shows no correlation with evolution in morphological characters available to paleontologists." The first of these is similar to Axelrod's ecospecies concept. The second is simply a statement of the fact that some species may shift the limits of their tolerance to environmental factors during geologic time.

Thermally anomalous fossil species may then have evolved physiologically but not morphologically. Such physiological evolution of species of course poses the greatest obstacle to interpretation of paleoclimates from distribution of species. A number of paleontologists have referred to this problem, but most feel that so few species have changed since late Tertiary time that the method is valid for upper Tertiary fossil faunas (Woodring, Bramlette, and Kew, 1946:102–103; Valentine, 1955:466; Valentine and Meade, 1961:1–2). Though the thermal implications of a few fossil species may be suspect, fossil faunas as a whole may accurately reflect environmental changes.

Still there is evidence that some "warm-water" taxa in the Elsmere Canyon fauna have changed their temperature tolerance limits with time. One of these is *Anadara; A. trilineata* has long been regarded by west coast paleontologists as a warm-water indicator because no living species of *Anadara* live north of Newport, California, and only one species, *Anadara multicostata,* lives north of Cedros Island (Keen, 1958:34). However, *Anadara trilineata* occurs in lower Pliocene rocks as far north as Coos Bay, Oregon. It is the only "warm-water" species in lower Pliocene rocks at Eureka, California, and with the exception of this species the entire fossil fauna suggests water colder than at present at that latitude (Faustman, 1964:108–110). Addicott (1970*b*:292) also has suggested that *A. trilineata* is not necessarily indicative of warm water, and Stanton and Dodd (1970:1099) have pointed out that late Cenozoic species of *Anadara* (and also *Arca*) occurred in the northwestern Pacific north of Japan. Clearly the genus *Anadara* (and also, apparently, *Arca*) has changed its temperature requirements since early Pliocene time.

Another illustration of change in temperature-distribution relationships is provided by *Dosinia jacalitosana.* This species apparently is closely related to *Dosinia ponderosa,* which lives today only south of Scammon's Lagoon, Lower California (Keen, 1958:136). It is the northernmost species in the genus. *Dosinia jacalitosana* occurs in the Elsmere Canyon fauna and in other Pliocene faunas as far north as Coalinga. It is associated at Elsmere Canyon with at least twenty molluscan species that do not live today as far south as Scammon's Lagoon (fig. 13) and at Coalinga with the temperate genera *Mya, Clinocardium, Patinopecten, Lunatia, Polytropa,* and *Terebratalia* (Adegoke, 1969:60–61, figs. 6, 6*a*), none of which live today as far south as Scammon's Lagoon. Stanton and Dodd (1970:1099) also have recognized the association of *Dosinia* with cool-water assemblages in Pliocene and Pleistocene rocks. Thus the genus *Dosinia* also seems to have changed its temperature requirements since early Pliocene time.

The thermal implications of the northern taxa in the Elsmere Canyon fauna are also ambiguous. It has been suggested by Stanton and Dodd (1970:1101) that both *Mya truncata,* a living species of the Oregonian province, and *Patinopecten lohri,* an extinct species related to temperate living taxa, are of questionable climatic value. *Mya truncata* and the other three species that live today in shallow water only in the Oregonian molluscan province must have been adapted to different thermal conditions in early Pliocene time. Early Pliocene shallow-water marine climate could not have been as cold as that in similar environments at Puget Sound today, especially in light of apparent evidence for warmer conditions.

Most of the Surian and Panamic genera are represented in the Elsmere Canyon fauna by extinct species. These genera have evolved morphologically since early Pliocene time, suggesting the possibility of evolutionary change in physiology. Possibly some of these species are descendants of truly warm-water species that ranged farther north than today in middle Tertiary time when the marine climate was warmer. Isolation of populations of such species in restricted coastal basins like the Ventura basin might have led to evolution of new species that were closely related to their warm-water ancestors but were adapted to colder-water conditions. Adegoke (1969:47) has suggested that these embayments favored the development of local endemic faunas with restricted geographic distribution.

Whatever the explanations for individual taxa, it is clear that some of the thermally anomalous species and genera in the Elsmere Canyon fauna live today under thermal conditions different from those under which they lived in the geologic past. Either these taxa have evolved physiologically or their distribution is independent of temperature.

TEMPERATURE INDEPENDENCE

The distribution of some thermally anomalous taxa may be independent of temperature. The evidence for independence of geographic distribution from thermal regime is discussed above. Such independence may explain the thermally anomalous occurrences of *Anadara, Dosinia,* and other genera that clearly lived under different thermal conditions in early Pliocene time than today.

ENVIRONMENTAL CHANGE

The fourth alternative is that changes in distribution of taxa during geologic time and the presence of thermally anomalous taxa in a fossil fauna do reflect changes in the environment, changes that can be reconstructed from the distribution of those fossil taxa. The presence of such taxa in fossil faunas has been interpreted in different ways by different paleontologists, depending on their understanding of the relationships between geographic distribution and the thermal requirements of marine poikilotherms. Most early studies were based on the assumption that distribution is limited by minimum and maximum survival temperatures. Authors of recent studies have more often shown an awareness of the more critical thermal requirements of many animals in other parts of the life cycle and have dealt with aspects of the thermal regime other than minimum and maximum temperatures. Past and present attempts to interpret early Pliocene climate in southern California are described below.

Summary of Previous Interpretations of Early Pliocene Climate

Paleoclimatic Studies of Pliocene Molluscan Faunas in Southern California

The climatic implications of the Elsmere Canyon fauna first were considered in Smith's (1919) study of west coast Tertiary climate. Of eighteen "southern" genera that Smith (1919:152–155) recognized in the lower Pliocene rocks of southern California, ten occur in the Elsmere Canyon fauna. The seventeen species in these genera are listed in table 4. Smith suggested that the presence of warm-water taxa in these rocks indicates that early Pliocene minimum winter temperatures were about 3°F warmer than the present-day minimum in this region. He thought that these conditions represented a stage in the gradual progressive cooling from early Tertiary to Pleistocene time (Smith, 1919:165–167).

TABLE 4

Smith's (1919:155) Southern Taxa That Occur in the Elsmere Canyon Fauna

Smith's taxon	Taxon in Elsmere Canyon fauna
Ficus nodiferus	*F. (Trophosycon) ocoyana*
Dosinia ponderosa	*D. jacalitosana*
Chione	*C. elsmerensis*
	C. fernandoensis
Lyropecten	*L. estrellanus*
Trachycardium	*T. quadragenarium*
Amiantis	*A. callosa*
Cypraea	*C. fernandoensis*
Cancellaria	*C. arnoldi*
	C. elsmerensis
	C. fernandoensis
	C. hamlini
	C. planospira
	C. rapa
	C. tritonidea
Dendraster ashleyi	*D. gibbsii*
Astrodapsis fernandoensis	*A. fernandoensis*

Smith's list of warm-water taxa in the Elsmere Canyon fauna (table 4) is evaluated as follows. *Trachycardium quadragenarium* and *Amiantis callosa* live today as far north as Santa Barbara, and species of *Cypraea* and *Dendraster* are as abundant today at Santa Barbara as they are in the Elsmere Canyon fauna. Since the genus *Astrodapsis* is extinct, its paleoclimatic significance is questionable. One species of *Cancellaria* lives as far north as Santa Barbara today (Burch, 1944–1946: no. 49:6), though the presence of several species in moderate abundance in the Elsmere Canyon fauna may reflect slightly warmer climate. Species of *Chione* are as abundant today at Santa Barbara as they are in the Elsmere Canyon fauna, but *Anomalocardia* is a Surian subgenus. Recent species of *Ficus, Lyropecten,* and *Dosinia* do not live north of Scammon's Lagoon (28° north latitude), though it is suggested above that the climatic significance of

Dosinia is doubtful. Of the "warm-water" taxa listed by Smith the only potentially valid thermally anomalous taxa occurring in Elsmere Canyon are ***Ficus***, *Lyropecten*, and *Chione* (*Anomalocardia*); several other thermally anomalous taxa listed by Smith occur in other lower Pliocene strata in southern California.

Woodring, Bramlette, and Kew (1946:101) and Valentine and Meade (1961:3) have pointed out that Smith consistently failed to consider northern extralimital species in favor of southern species where they occur in the same fauna and also that he did not consider species that do not live today where the extralimital species live. It is also important to realize that Smith's interpretation of minimum winter temperatures rests on the assumption that distribution is controlled by temperature effects on survival.

Schenck and Keen (1936, 1937) developed a controversial paleoclimatic indicator called the median of midpoints-of-ranges index. This index is the median of the Recent range midpoints of all the living species in a fossil fauna. The present-day thermal regime at the index latitude should be the same as the thermal regime in which the fossil fauna lived. The median of midpoints for the Elsmere Canyon fauna was calculated by Schenck and Keen (1937:167–168, fig. 3) at approximately 34° 45′ north latitude on the basis of the faunal list published by English (1914). I have recalculated the index, using the larger faunal list now available. The ranges and midpoints of ranges of all the Recent species in the Elsmere Canyon fauna are plotted in figure 14 (in pocket), and the median of midpoints is calculated to be 35° 54′, about 1.5° north of the latitude of Elsmere Canyon. This result is consistent with the interpretation that early Pliocene climate at Elsmere Canyon was not substantially different from the climate at the same latitude today.

The midpoints-of-ranges index has been shown by Newell (1948:155–156, 165) to be invalid for defining limits of marine zoogeographic provinces. His criticism, however, was not directed at the use of midpoints in interpreting paleoclimates, and he admitted the theoretical validity of the use of range midpoints for such interpretations.

Valentine and Meade (1961:5) have pointed out other weaknesses in the midpoints-of-ranges index. This method assumes that latitudinal midpoints of species' ranges are critical localities and that temperatures at midpoints are critical. A further difficulty is posed by the various kinds of relationships between environmental temperature and distribution of marine poikilotherms, as discussed above; range end points have different thermal significance for different species. They also pointed out that this method does not allow for complex changes in thermal regime: changes in temperature ranges, in seasonal temperature distribution, or in other aspects of the thermal regime.

Yet indexes calculated by Schenck and Keen (1937:165, fig. 2) for collections of shallow-water Recent molluscs in southern California and in Baja California were all within 2 degrees of latitude of their actual locations, suggesting that the midpoints-of-ranges index may have some value. Farther north, on the other hand, the error is larger.

Durham (1950, 1954) calculated the positions of February marine isotherms along the west coast of North America during Tertiary time, basing his interpretation on many of the known fossil megafaunas in this region. The Elsmere Canyon fauna was not discussed in these papers, but the presence of tropical molluscan genera in several other lower Pliocene formations was interpreted as reflecting substantially warmer early Pliocene winter temperatures in southern California. Durham's (1950: fig. 3) chart of past and present positions of February marine isotherms shows a minimum temperature of about 17°C for the latitude of Elsmere Canyon in early Pliocene time, compared with 13° at that latitude today. In the later paper (Durham, 1954: fig. 4) a chart of past positions of February isotherms in the Los Angeles area shows a decrease since early

TABLE 5

WINTERER AND DURHAM'S (1962:308, TABLE 5) TOWSLEY FORMATION SHALLOW-WATER MOLLUSCAN SPECIES NOT LIVING AT THE SAME LATITUDE TODAY

Species or Recent relatives living south of Santa Barbara:
Bittium cf. *B. asperum* (Gabb)
Miltha xantusi (Dall)
Lucina excavata (Carpenter)
Spisula hemphilli (Dall)
Dosinia ponderosa (Gray)
Amiantis callosa (Conrad)
Species living north of Santa Barbara:
Mya truncata Linnaeus

Pliocene time of 3°C, from 18° to 15°. More recent temperature data (U.S. Navy, 1956: charts 14, 24) indicate a minimum mean monthly temperature at this latitude today of 12°C (fig. 16).

Winterer and Durham (1962:308, table 5) listed as occurring in the Towsley Formation nine shallow-water species that do not live at the same latitude today. Seven of these species occur in the Elsmere Canyon fauna and are listed in table 5. Winterer and Durham (1962:308) interpreted the Towsley fauna as indicating shallow-water temperatures several degrees warmer than at Ventura today.

The seven species in table 5 are evaluated as follows. *Amiantis callosa* ranges north to Santa Barbara. *Lucina excavata* [= *Here excavata*], *Spisula hemphilli,* and *Bittium asperum* live as far north as San Pedro. *Miltha xantusi* does not live north of 24° north latitude today, but it is shown below that the thermal implications of this species are doubtful. The same is suggested above for *Dosinia* and *Mya truncata.* Two other extralimital species recorded by Winterer and Durham (1962:308, table 5) do not occur at Elsmere Canyon; *Turritella* cf. *T. gonostoma hemphilli* Merriam and *Ostrea vespertina* Conrad may be part of a small warm-water element in the Towsley Formation.

Addicott (1970*b*:304–305, table 2, fig. 7) also has interpreted the Elsmere Canyon fauna as indicating a marine climate substantially warmer than today at that latitude, consistent with similar interpretations for lower Pliocene faunas in central California.

PALEOCLIMATIC STUDIES OF OTHER MOLLUSCAN FAUNAS

The upper Miocene Castaic Formation, probably only slightly older than the Elsmere Canyon section of the Towsley Formation, was deposited in the Ventura basin between 20 and 40 miles northwest of the Elsmere Canyon area. These two areas have since been brought closer together by lateral movement on the San Gabriel fault (Crowell, 1962:39). The Castaic fauna was interpreted by Stanton (1966:25) to indicate a minimum February surface temperature of 19° to 20°C, substantially higher than today at that latitude. Although the two faunas are similar (twenty-seven of the fifty-four definitely identified Castaic species occur in Elsmere Canyon), the only Castaic warm-water taxa that occur in Elsmere Canyon are *Miltha xantusi, Dosinia* sp., and *Ficus ocoyana.* Stanton (1966: fig. 2) listed these additional warm-water taxa from the Castaic Formation: *Glycymeris* cf. *G. gigantea* (Reeve), *Spondylus* sp., *Eucrassatella* (*Hybolophus*) *subgibbosa* (Hanna), *Liotia carinata* Carpenter, *Nerita* sp., *Trochita* cf. *T. trochiformis* Born, *Polinices uber* (Valenciennes), *Oliva spicata* (Röding), *Marginella* cf. *M. albuminosa* Dall, and *Conus* sp. The warm-water element in the Ventura basin seems to have been better developed in late Miocene than in early Pliocene time.

Lower Pliocene rocks are not well exposed in the Los Angeles basin (fig. 1), so little is known of their fauna. Because nearly all the twenty-six molluscan species listed by Woodring (1938:11) are deep-water species, no temperature inferences have been drawn from this fauna. No shallow-water faunas of early Pliocene age are known from the Los Angeles basin.

Lower Pliocene fossil faunas in the Coalinga region have been recorded by Arnold (1909: 24–25), Nomland (1917:211–213), and Adegoke (1969: figs. 6, 6*a*). Adegoke (p. 53) suggested that tropical to subtropical marine climates in middle and late Miocene time in the Coalinga region were followed by progressive cooling until late Pliocene time. He thought that several warm-water genera in the lower Etchegoin Formation, including *Anadara, Chione, Lyropecten, Dosinia, Psammotreta, Trachycardium, Forreria,* and *Turritella,* probably indicate minimum surface temperatures of 13° to 14°C during Pliocene time (*ibid.,* pp. 60–61).

Stanton and Dodd (1970:1102–1103, text fig. 5) interpreted the Jacalitos (lower Etchegoin of Adegoke, 1969) fauna as indicating a shallow-water marine climate like that today at latitude 30° to 31°, some 6 to 7 degrees south of the Coalinga region. This interpretation was based on comparison with the provincial locations of modern faunas. Of ten extralimital genera on their list for the Jacalitos Formation, however, only four—*Anomalocardia, Lyropecten, Miltha,* and *Dosinia*—occur today only south of latitude 31°, and the latter two species are of questionable climatic significance. The other six genera all range at least as far north as San Pedro at latitude 34°. Like the Elsmere Canyon fauna, however, the lower Etchegoin fauna has a small component of southern taxa that may reflect slightly warmer conditions in early Pliocene time.

According to Durham and Addicott (1965:A18–A19) the Pancho Rico fauna in the Salinas Valley (fig. 1) suggests a substantially warmer marine climate at that latitude in early Pliocene time than at present. Molluscan taxa with affinities to the Californian molluscan province (Cedros Island to Point Conception) include *Crucibulum (Dispotaea)*, *Anadara, Lyropecten,* and a *Chione* allied to *C. gnidia.* Taxa more suggestive of the Surian province include *Diodora* sp., *Turritella gonostoma hemphilli, T. vanvlecki, T.* sp., *Crassispira* sp., *Dosinia, Arca,* a *Glans* comparable to the living *G. radiata,* and *Miltha* sp. On the basis of Durham and Addicott's (1965: table 1) faunal list Stanton and Dodd (1970:1102, fig. 5) suggested a shallow-water climate for the Pancho Rico fauna similar to that at latitude 30° to 31°, about 6 to 7 degrees south of the Salinas Valley. Addicott (1970*b:* table 1) later indicated that thirteen genera from the Pancho Rico Formation today are restricted to latitudes south of Cedros Island. These genera include *Anomalocardia* and *Turricula* in addition to most of the taxa listed above, but *Anadara, Dosinia, Arca,* and *Miltha* are shown here to be of doubtful climatic significance. Several other genera, including *Diodora* and *Crassispira* (McLean, 1969:13, 54), live as far north as southern California. Thus there seem to be fewer than thirteen valid thermally anomalous taxa in the Pancho Rico Formation, though a significant southern element does exist in that fauna. A smaller and less abundant northern element in the Pancho Rico fauna includes *Chlamys* aff. *C. nipponensis, Mya arenaria, M. truncata, Tellina* cf. *T. lutea, Macoma brota,* and *Siliqua* cf. *S. media* (Durham and Addicott, 1965:A19).

Of the sixteen Recent species in the fauna of the lower Pliocene Tahana Sandstone Member and Pomponio Sandy Mudstone Member of the Purisima Formation (Cummings et al., 1962: pl. 24) in the Santa Cruz Mountains (fig. 1), the geographic ranges of all but two overlap at that same latitude today. *Macoma brota lipara* (Dall) lives only north of Puget Sound, and *Yoldia thraciaeformis* (Storer) lives from Oregon north. The wide disparity between the ranges of these species and those of the rest of the fauna suggests that the extralimital species have changed ecologically or that their distribution is independent of temperature. The remainder of the fauna suggests early Pliocene climate similar to that at the same latitude today.

A molluscan fauna of thirty-seven species from the Tahana Member in Portola Valley includes several southern taxa, among them *?Calicantharus portolaensis, Anadara trilineata, ?Clementia* sp., *Dosinia ponderosa,* and *Psammotreta biangulata* (Addicott, 1969*b:* table 4). The climatic significance of *Anadara* and *Dosinia* has been questioned, and the identification of *Calicantharus* and *Clementia* is uncertain, but *Psammotreta biangulata* lives only south of Point Conception, three degrees of latitude south of its occurrence in the Purisima Formation. Thus there seems to be a small southern element in the lower Pliocene rocks of the Santa Cruz Mountains, though this element apparently is not so well developed as it is in rocks of the same age farther south. Durham and Addicott (1965:A19) also have commented on the markedly cooler water in this region relative to more southern localities in early Pliocene time.

Using Schenck and Keen's (1937) median of midpoints-of-ranges index, Faustman (1964:108–110) calculated that marine water temperatures in the Eel River basin (fig. 1) apparently were several degrees cooler in middle Pliocene time than at present. This index could not be used with the lower Pliocene Pullen and Eel River formations because of the small number of species in the megafauna. Rather, the presence in the Pullen Formation of abundant *Anadara trilineata* was taken to indicate water warmer than at present in early Pliocene time. Faustman (1964: 109–110) also said that Ogle (1953) "reported that the foraminiferal fauna is similar to that of the Repetto Formation in southern California" and concluded that "this may indicate that warm-temperature surface conditions continued to exist at the time of deposition of the Eel River Formation." Ogle (1953:47), however, merely correlated the age of the upper part of the Pullen Formation and of all of the Eel River Formation with not only the Repetto Formation, but also the Jacalitos (lower Etchegoin of Adegoke, 1969) and Purisima formations. He made no reference to environmental similarities. With the exception of *Anadara trilineata,* which is shown above to be of questionable paleoclimatic significance, the lower and middle Pliocene molluscan fauna in the Eel River basin suggests water colder than at present at that latitude (Faustman, 1964:108–110). This interpretation, however, is in opposition to the climatic inferences of planktonic foraminifers, which have been shown by Ingle (1968:147) to indicate substantially warmer early Pliocene temperatures in this region.

In a series of recent studies of Tertiary paleoclimatic trends based on molluscan faunas, Addicott (1968; 1969*a;* 1970*a:* fig. 2; 1970*b*) has shown that climate in the northeastern Pacific Ocean was substantially warmer in early Pliocene time than it is today. These studies differ from those of earlier molluscan workers (but are similar to Ingle's previous foraminifer studies described below) in that they indicate a Miocene anomaly in the progressive cooling that occurred during the Tertiary Period. Addicott has suggested that temperatures began to increase in middle Oligocene time, reached a peak in middle to late Miocene time, and declined rapidly to lower than Oligocene levels by early to middle Pliocene time.

PLANKTONIC FORAMINIFERAL EVIDENCE FOR EARLY PLIOCENE CLIMATE IN SOUTHERN CALIFORNIA

Ingle (1967:315, fig. 36) has shown that planktonic foraminiferal trends also suggest an overall decrease in surface temperature modified by a series of large oscillations during Tertiary time. One of these oscillations occurred in late Miocene and early Pliocene time, when the cool-temperate faunas that have prevailed in southern California since late middle Miocene time were briefly replaced by subtropical faunas (Ingle, 1967: 284–285). This trend is reflected by dextral coiling of *Globigerina pachyderma* and the presence of a number of tropical and warm-temperate species of foraminifers in upper Miocene and lower Pliocene rocks in the Los Angeles and Ventura basins. Ingle (1967:298) has said that sea surface temperatures at latitude 33° north may have exceeded 25°C during early Pliocene time, and he has suggested an annual range of temperature of about 16° to about 22°C at the Miocene-Pliocene boundary, compared with about 15° to about 18° at that latitude today. This interpretation differs from those of earlier workers in that high early Pliocene temperatures were thought by Ingle to reflect a warming trend in late Miocene and early Pliocene time, while others have regarded these conditions as a stage in the progressive cooling of Tertiary marine climate.

Ingle (1967:290, fig. 30) placed the Miocene-Pliocene boundary just above the base of the Repetto Formation in the Los Angeles basin, and he apparently accepted Winterer and Durham's (1962:294) placement of the boundary in the eastern Ventura basin. Ingle's paleoclimatic interpretation thus suggests that the lower Pliocene Elsmere Canyon fauna lived in a thermal regime that was several degrees warmer all year than the thermal regime at that latitude today.

Ingle's and Addicott's interpretations of Tertiary climatic trends, respectively based on foraminiferal and molluscan evidence, differ in their placement on the time scale. According to Ingle (1967: fig. 36) the high-temperature peak occurred in early to middle Pliocene time; according to Addicott (1969*a:* figs. 1, 3; 1970*a:* fig. 7), in middle to late Miocene time. Though the thermal implications for the Elsmere Canyon fauna are the same, high early Pliocene temperatures represent a stage in the warming trend in Ingle's interpretation, and a stage in the succeeding cooling trend in Addicott's. Clearly the paleoclimatic inferences of foraminifers and molluscs are contradictory in Miocene and Pliocene strata of California.

Ingle (1967:278–291) has suggested that changes in the thermal regime off southern California during Cenozoic time may have been caused by refrigeration and warming at the poles and consequent changes in temperature of ocean currents. He pointed out that the southern California region today is the location of a transition zone between areas dominated by subarctic and by warm-temperate planktonic assemblages and that the position of this transition zone is controlled by the south-flowing cold-water California Current. The subarctic assemblage extends south to latitude 25° north today and has existed at least as far south as latitude 30° north since middle Miocene time. The tropical assemblage that is restricted south of latitude 25° in these waters lives as far north as latitude 43° in the western Pacific warm-water Kuroshio Current. Clearly, planktonic foraminiferal assemblages in the eastern Pacific are strongly influenced by the California Current, and changes in this current during Cenozoic time probably have been responsible for changes in distribution of these assemblages. Ingle (1967: 291) suggested that these changes in foraminiferal assemblages and in marine thermal regime in southern California probably reflect changes in the Arctic region such as the advance and recession of glaciers and sea ice.

The possible role of changes in the California Current had been recognized earlier by Gale (in Grant and Gale, 1931:39, 64) and Woodring, Bramlette, and Kew (1946:102), who suggested that the Channel Islands peninsula may have kept south-flowing cold water out of the Los Angeles basin in late Pleistocene time, allowing southern species to extend their ranges to the north. This expansion of warm water to the north would not by itself explain the mixing of warm- and cold-water species, but such changes in coastal geography may have played a role in faunal redistribution.

EARLY PLIOCENE MARINE CLIMATE IN NEW ZEALAND

A number of recent studies of Tertiary marine climatic trends in New Zealand have been based on fossil foraminifers (Lewis, 1968; Jenkins, 1968), corals (Keyes, 1968), molluscs (Beu and Maxwell, 1968), and other organisms (Hornibrook, 1967). The first four of these studies suggest a high-temperature peak between late Eocene and early Miocene time followed by progressive cooling well into the Pliocene Epoch. Three of these studies suggest that early Pliocene temperatures were equal to or less than present-day temperatures at the same latitudes, but the fourth suggests somewhat higher and fluctuating early Pliocene temperatures. Hornibrook (1967:37) also suggested that New Zealand Pliocene marine temperatures were about the same as at present or a little warmer. Oxygen isotope analyses of New Zealand Tertiary fossils by Devereux (1967:1004, fig. 1) indicate a high-temperature peak in middle Miocene time followed by progressive cooling into the Pleistocene Epoch with early Pliocene temperatures the same as today's at the same latitudes. Thus most of the available paleontologic and geochemical evidence suggests that early Pliocene marine temperatures in New Zealand were very similar to those at the same latitudes today.

EARLY PLIOCENE TERRESTRIAL CLIMATE IN SOUTHERN CALIFORNIA

The first eastern North Pacific paleoclimatic studies to analyze aspects of the thermal regime more complex than simple minimum and maximum temperatures were made by paleobotanists. Axelrod (1941:549; 1967*b*:6–8) suggested that changes in distribution of fossil plant species in western North America during late Cenozoic time reflect decreasing equability, or increasing annual range of temperature, in terrestrial climate beginning after early Pliocene time.

Associations of tropical with nontropical plants and animals occur today only in areas of high equability, such as the Mexican cloud forest and the central African plateau (Axelrod, 1967*b*:7–11, 24). In such areas the annual temperature range is small enough so that minimum temperatures are not too low for tropical species and maximum temperatures are not too high for nontropical species. Similar conditions may have existed in the geologic past, accounting for associations of fossil floral and faunal elements that live today in different environments.

Middle Tertiary floras of California included a number of species that now live in northern and central Mexico (Axelrod, 1956:257–270). The presence of these subtropical taxa in Miocene and Pliocene rocks in California probably indicates a moister, more equable climate there at that time (Axelrod, 1967*a*:298–308). Axelrod (oral communication, 1968) believes that the general cooling trend of the early Tertiary Period had leveled off by middle Miocene time and that later climatic change brought a reduction in summer precipitation and an increase in the annual range of temperature. In other words, equability was reduced, while warmth of climate changed relatively little. The restriction of the subtropical floral element to the south was a result both of the reduced summer rain and of the lower winter temperatures. The terrestrial plant species that were affected must be limited geographically by survival temperatures.

A similar reduction in equability, though in this instance accompanied by a decrease in climatic warmth, was suggested by Dorf (in Schwarzbach, 1963:170) from the evidence of fossil floras in southeastern Oregon. He estimated an increase in annual range of temperature since early Pliocene time from 9.4°C (8.6° to 18°) to 24.5°C (–2.5° to 22°) and a decrease in mean annual temperature from 14.2° to 10°C.

Wolfe and Hopkins (1967:74), interpreting lower Pliocene fossil floras in northwestern North America as indicating that terrestrial climate had become essentially like that of today, suggested that slight cooling probably occurred during the Pliocene Epoch.

The widespread extinctions of large mammals during Quaternary time also have been attributed by Axelrod (1967*b*:24–25) to decreased equability. He suggested that tropical species of the rich Tertiary mammalian faunas of North America and Eurasia were eliminated from those continents by reduced minimum temperatures during glacial episodes, when climatic extremes were greater.

Thus the evidence from terrestrial fossil biotas suggests that early Pliocene terrestrial climate was not in general warmer or cooler than climate today in western North America, but it was more equable. That is, there was a narrower annual range of temperature. Winters were mild enough that southern species could live farther north than today, and summers were cool enough that northern species could live farther south. This interpretation rests on the assumption that geographic ranges of most terrestrial mammal and plant species are limited by temperatures required for survival, winter temperatures toward the north and summer temperatures toward the south.

COMPARISON OF TERRESTRIAL WITH MARINE CLIMATE

Before comparing these interpretations with possible conditions in the sea, it must be pointed out that the paleoecological significance of temperature changes may be different in marine and terrestrial environments. Although broad thermal patterns must correspond in the two environments, local conditions may be quite different, owing to irregularities in topography and in atmospheric and oceanic circulation. Further, because of the narrow range of temperature in the sea, any climatic change corresponding to a similar change on land would be of much smaller magnitude and probably would have less effect on the fauna.

The narrow range of temperature in the sea also has important implications for the physiological relationships of organisms to environmental temperature. The minimum temperature of –2.5°C in the sea is much higher than the minimum of –88:3° recorded on land (Kinne, 1970:330). At temperatures that low, terrestrial organisms may be killed by freezing. Animals cannot live in northern latitudes unless they can maintain a high body temperature through physiological or morphological temperature controls or can survive the cold season in an inactive state. Body temperatures of marine poikilotherms are about the same as the temperature of the water in which they live. As long as this medium is fluid, these animals cannot freeze, and many species are adapted to live in even the coldest water. The narrow temperature range of the marine environment also limits the possible magnitude of temperature changes. Thus marine organisms are not subjected to the extreme range of temperature possible on land. Though marine animals are killed by sudden temperature changes (Brongersma-Sanders, 1957:947–951), and their distribution is probably in part determined by minimum and maximum survival temperatures, it is probable that effects of temperature on reproduction and other phases of the life cycle are important factors limiting the distribution of many species. This difference in temperature ranges in the two environments also may account for the fact that survival temperatures appear to be more critical in limiting distribution on the land than in the sea.

Paleoclimatic Interpretation of the Elsmere Canyon Fauna

POSSIBLE ROLE OF LOCAL TEMPORARY THERMAL CHANGES

Woodring, Bramlette, and Kew (1946:102) suggested that local temporary changes in temperature may result in temporary small extensions or reductions in ranges of marine poikilotherms, but apparently they did not feel that this could account for many of the anomalous associations in Tertiary faunas. A somewhat similar hypothesis has been proposed by Valentine (1955:466–468) and endorsed by Emerson (1956:325) and Valentine and Meade (1961:6). They have suggested that the mixing in fossil faunas of thermally anomalous species was due to the nearly contemporaneous distribution of water both warmer and colder than is normal today. Present-day northern species extended their ranges to the south in areas of upwelling cold water along the outer coast, while present-day southern species extended their ranges to the north in sheltered waters. This hypothesis is supported by the common occurrence in west coast Pleistocene rocks of cold-water fossil species in exposed-coast faunas and warm-water species in sheltered-water faunas (Valentine, 1955:464–465; 1957:302; 1960:161; Addicott and Emerson, 1959:3; Emerson and Chace, 1959:340). There are exceptions to this pattern. For example, *Thais biserialis* (Blainville), which lives today from Cedros Island south (Keen, 1958:372), occurs in the exposed-coast Pleistocene fauna at Punta China, just south of Ensenada (Emerson, 1956:332). *Chione picta* Willett, which lives today from Magdalena Bay south (Keen, 1958:146), occurs in the exposed-coast Pleistocene fauna at Punta Descanso, north of Ensenada (Valentine, 1957:296). Exposed-coast faunas of upper Pleistocene terrace deposits in the Palos Verdes Hills at Los Angeles contain a number of both northern and southern species in the same beds (Woodring, Bramlette, and Kew, 1946:87–89). Thus the mixing of species from cold upwelling water and heated shallow embayments may explain some but probably does not explain all thermally anomalous species in fossil faunas.

The Elsmere Canyon fauna contains both northern and southern extralimital species in the same beds, suggesting their presence is not due to contemporaneous heating of shallow water and upwelling of cold water. Further, the four Recent cold-water species in the Elsmere Canyon fauna today live in shallow water only north of Puget Sound, 14 degrees of latitude to the north. Upwelling water cold enough for these four northern species would be expected to have had more pronounced effects on the overall composition of the fauna. Also, though *Natica clausa, Fusitriton oregonensis,* and *Neptunea lyrata* all live today in deep water off southern or central California, apparently none of them occurs there in shallow upwelling water. As is shown above, however, the northern extralimital species in this fauna are of questionable climatic significance; probably these species have adapted to different thermal conditions. Thus the role of solar heating of shallow water cannot be totally disregarded as a possible factor in the changed distribution of the southern extralimital species.

TEMPERATURE-DISTRIBUTION RELATIONSHIPS AND COMPLEX CHANGES IN THERMAL REGIME

Most of the paleoclimatic studies of lower Pliocene faunas described above are founded on two basic assumptions. One assumption is that all organisms are limited geographically by survival temperatures, for nearly all these interpretations deal with minimum temperatures at northern range end points. The second assumption is that Tertiary climatic change meant only simple general warming or cooling of climate: increase or decrease in seasonal minimum, seasonal maximum, or average yearly temperature. A few of these studies do not require both assumptions. Schenck and Keen (1936, 1937) did not specify minimum or maximum temperatures, so they were not necessarily dealing with survival temperatures alone, but their conclusions imply general warming or cooling of climate. Ingle (1967) suggested changes in annual range of temperature, though he did not specifically state that such changes are important in limiting distribution of species. Axelrod (1941, 1956, 1967*a,* 1967*b*) suggested changes in annual range of temperature but assumed that terrestrial organisms are limited by survival temperatures. Several other recent studies have questioned the validity of these assumptions for marine poikilotherms and have recognized more complex changes in thermal regime.

POSSIBLE ROLE OF TEMPERATURE REQUIREMENTS FOR SPAWNING AND JUVENILE GROWTH

Valentine (1955:467) found that in regions having a broad annual range of temperature southern species that breed in summer are sometimes associated with northern species that breed in winter. This conclusion implies that northern range end points of southern species are determined by summer rather than winter temperatures and that southern range end points of northern species are determined by winter rather than summer temperatures (see above discussion of temperature-distribution relationships). Valentine (1955:467) and Valentine and Meade (1961:6) have suggested that similar differences in annual range of temperature (reduced equability) and other thermal variables in the past have been responsible for thermally anomalous fossil faunas.

At least one southern species in the Elsmere Canyon fauna clearly is not geographically restricted by minimum temperature. *Miltha xantusi* often has been taken to indicate the tropical aspect of fossil faunas (Smith, 1919:153; Durham, 1950: 1248; Winterer and Durham, 1962:308; Addicott and Vedder, 1963:63–64; Stanton, 1966:25). This species is known in the Recent fauna off Cape San Lucas, Lower California, at 23° north latitude. It has been reported in the intertidal zone (Pilsbry and Lowe, 1933:137) but is most abundant in 30 or more fathoms (Keen, 1958:98; Vokes, 1969:94). Winter water temperatures in December 1954 and January 1959, at depths of 36 and 44 fathoms, respectively, off Cape San Lucas were very close to surface water temperatures off Santa Barbara at the same times (Scripps Institution of Oceanography, 1965: vol. 1954, CCOFI cruise 5412; vol. 1959, CCOFI cruise 5901) (table 6). While these data are limited, they at least suggest that minimum water temperatures in these two places may be comparable.

Miltha xantusi is abundant in Elsmere Canyon in only a few localities, which were probably in water less than 10 fathoms deep. Apparently this species lived in very shallow water in the early Pliocene Ventura basin, water that may have been very little colder than the deeper water in which it now lives off Cape San Lucas. Its restricted Recent geographic range, therefore, apparently reflects the effects on its distribution of some factor other than minimum temperature.

TABLE 6

COMPARISON OF RECENT SURFACE-WATER MARINE TEMPERATURES AT SANTA BARBARA AND DEEPWATER TEMPERATURES AT CAPE SAN LUCAS

Date	CCOFI cruise	Santa Barbara station 82.49		Cape San Lucas station 153.16	
		Depth (m)	Temperature (°C)	Depth (m)	Temperature (°C)
December 1954	5412	0	15.90	0 71 107	24.50 15.80 13.49
January 1959	5901	0	15.40	0 49 73 99	22.63 19.38 16.78 14.60

NOTE: Data from Scripps Institution of Oceanography (1965).

The same conclusion is suggested by the distribution of Recent Surian and Panamic taxa that lived in the late Tertiary Ventura basin. Figure 17 compares the Recent latitudinal ranges in the Gulf of California and on the outer coast of Lower California of ten southern molluscan genera that occur in the Castaic and Towsley formations. The ranges of seven of these ten genera extend from 3 to 8 degrees of latitude farther north in the gulf than on the outer coast. It can be seen from figure 16 that winter water temperatures are as cold at the north end of the gulf as on the outer coast at the same latitude and are colder by several degrees than those on the southern half of the outer coast. Clearly the northern range end points of these taxa are not determined by minimum winter temperatures.

Figure 15 shows that summer water temperatures are several degrees higher at the north end of the Gulf of California than on the outer coast at the same latitude. Thus seven of ten southern genera (fig. 17) live to the north of their outer coastal range end points in water that is colder in winter and warmer in summer. Their northern range end points may then be determined not by winter cold but by lack of summer warmth, and the post–early Pliocene restriction of the Surian-Panamic faunal element several degrees of latitude south of the Ventura basin may have been the result of a decrease in summer temperature and in annual range of temperature (increase in equability) rather than of a decrease in winter temperature.

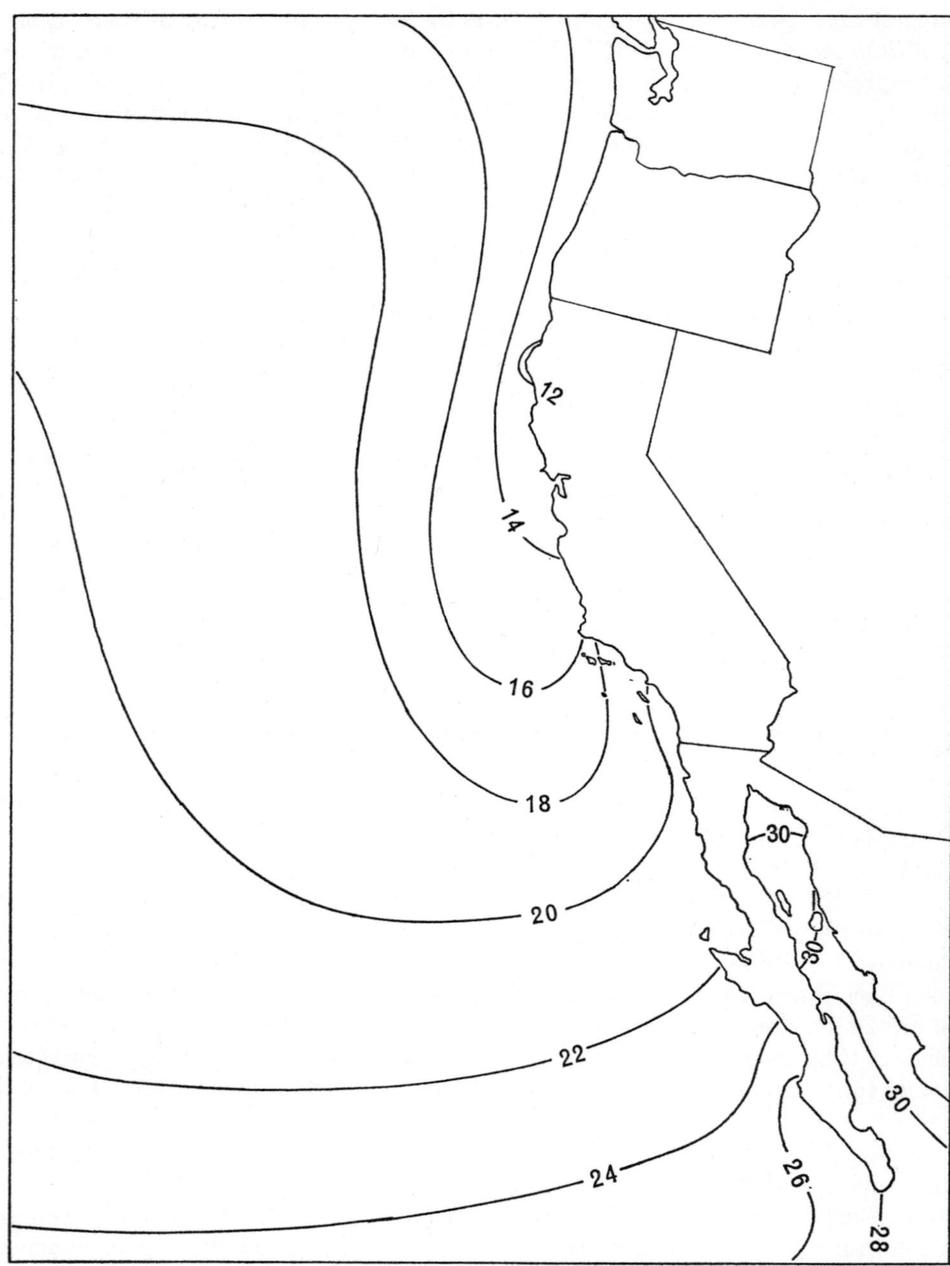

Fig. 15. August surface isotherms in °C (Ricketts and Calvin, 1939:380; U.S. Navy, 1956: charts 90, 100; Roden, 1964:46, fig. 7).

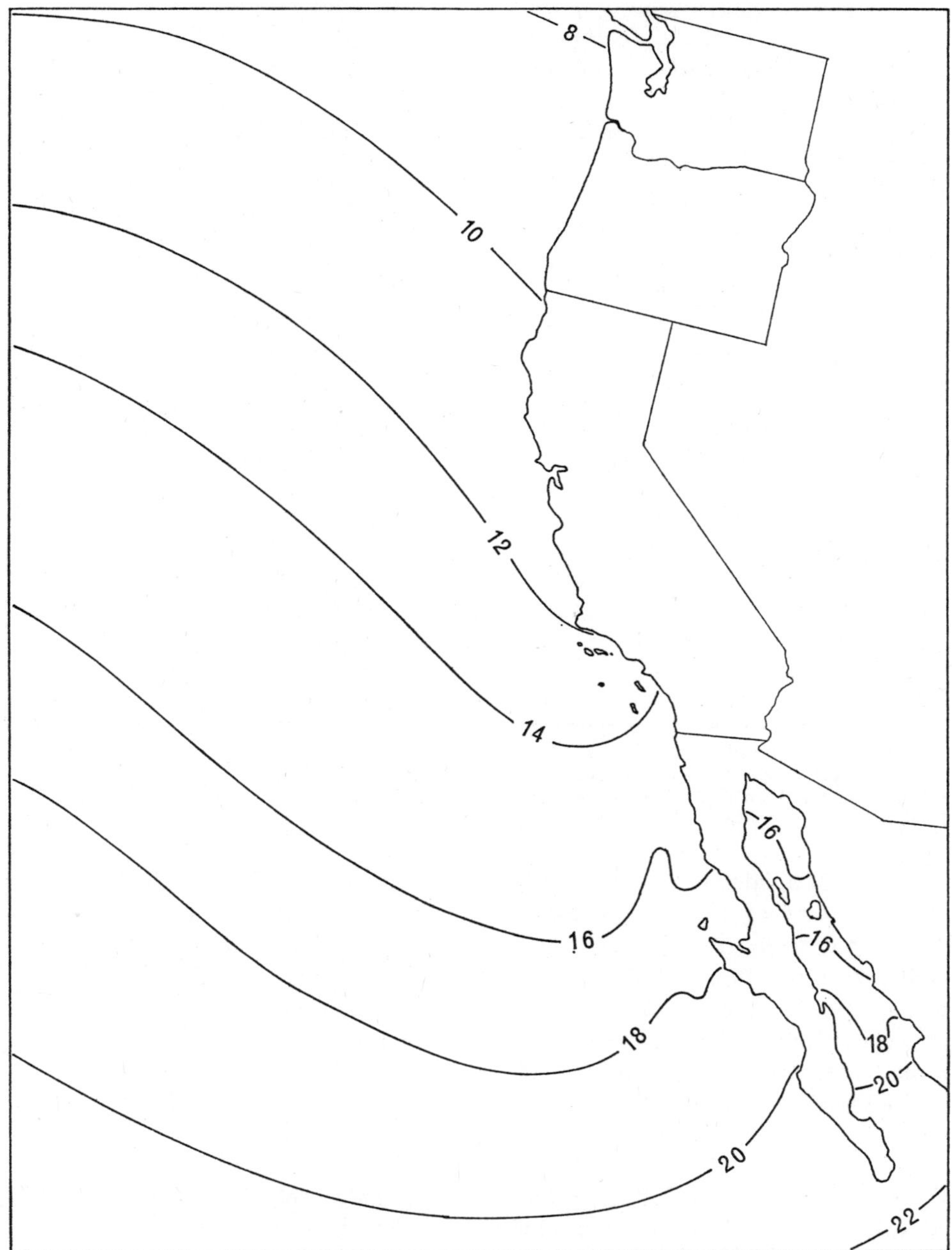

Fig. 16. February surface isotherms in °C (Ricketts and Calvin, 1939:381; U.S. Navy, 1956: charts 14, 24; Roden, 1964:46, fig. 7).

A similar interpretation has been suggested by Hedgpeth (1957:373–376), who postulated that a critical factor in limiting distribution of shallow-water marine molluscs may be the length of the season during which temperatures are high enough for reproduction and early growth. Hall (1964:227–228) found that many of the genera considered tropical by Durham (1950:1248) include species that live in water colder than 14°C for part of the year, but that nearly all the species in these genera live in water that is 20°C for at least four months of the year. The minimum temperature for these genera is 10°C. These taxa represent Hall's (1964:

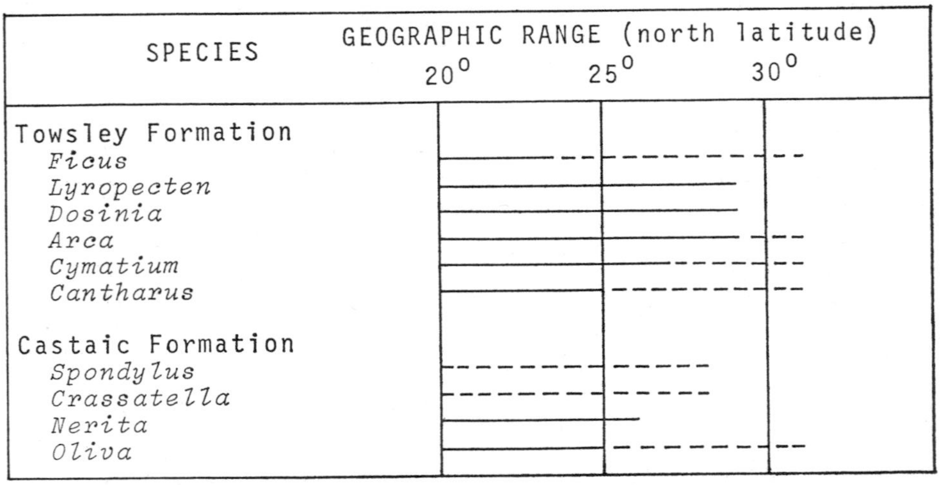

SPECIES

GEOGRAPHIC RANGE (north latitude)

20° 25° 30°

Towsley Formation

Ficus

Lyropecten

Dosinia

Arca

Cymatium

Cantharus

Castaic Formation

Spondylus

Crassatella

Nerita

Oliva

Fig. 17. Geographic distribution of Recent species of tropical genera that occur in the Towsley and Castaic formations. Ranges on the outer coast of Lower California are shown by solid lines; northern extensions of ranges in the Gulf of California are shown by dashed lines. Data from Keen (1958).

table 11) Magdalenan molluscan province (Surian province of Valentine), from Cape San Lucas to Scammon's Lagoon (23° to 28° north latitude). His Californian province extends from Scammon's Lagoon to Point Conception (34° north latitude). This province includes molluscan species that live in water that is 18°C for at least four months, with a minimum of 10°C. The duration of summer warmth may be a more critical factor in determining molluscan distribution than is minimum temperature. Southern extralimital species in fossil assemblages may then be limited to the north by requirements for high summer temperatures for breeding.

TEMPERATENESS AND EFFECTIVE TEMPERATURE

Clearly there are several divergent views among paleontologists as to late Cenozoic climatic history. Changes in faunal and floral distribution have traditionally been regarded as indicative of general warming or cooling of climate and have been defined by geographic shifting of isotherms. But some paleobotanical evidence suggests that thermally anomalous fossil biotas have been produced by more equable climates in the past, and the distribution of some living marine poikilotherms suggests that thermally anomalous fossil biotas have been produced by less equable climates. Two concepts that have been found by botanists to be useful in understanding the effects of the overall thermal regime on plant and animal distribution have been employed below in an attempt to evaluate the paleoclimatic implications of the Elsmere Canyon fauna. These climatic concepts—effective temperature and temperateness—have been described by Bailey (1960, 1964) and Axelrod (1967*b*:5–11).

Warmth of climate is measured by Bailey's (1960:1–10) effective temperature, which is a function of the average temperature of the warmest and coldest months of the year. Effective temperature is a measure of seasonal, rather than annual, warmth, since the scale by which it is measured takes into account the duration of the warm part of the year. Graphic and mathematical methods of calculation of effective temperature are shown in figure 18.

Equability is measured by Bailey's (1960:10–13) temperateness, which is a function of the average annual temperature and the average annual range of temperature. Graphic and mathematical methods of calculation of the temperateness index are shown in figure 19. A constant temperature of 14°C is assigned an index of 100. This temperature is close to the mean temperature of the earth and also lies midway between the thermal limits of tropical and polar climates (Bailey, 1960: 10). The greater the deviation of temperature from 14° and the larger the seasonal fluctuation of temperature, the lower the temperateness index (fig. 19). A lower index means a less equable climate.

EARLY PLIOCENE TEMPERATENESS AND EFFECTIVE TEMPERATURE IN THE VENTURA BASIN

The Recent Surian-Panamic molluscan fauna is restricted, on the outer coast of Lower California, south of Cedros Island (28° north latitude). The annual range of monthly mean surface-water temperatures (not mean annual range of temperature) is approximately 17° to 22°C (table 7) with four months at 20°C (Hall, 1964: fig. 2, table 11; my figs. 15, 16). The effective temperature, calculated from figure 18, is 17.4°C. Estimating the mean annual temperature (for which data are not available) at 19° to 20°C, and the mean annual range of temperature at 7° to 8°C, the temperateness index, calculated from figure 19, is close to 68.

The same fauna extends to the northern end of the Gulf of California, and the same data from this region are also summarized in table 7. Mean annual temperature and mean annual range of temperature are estimated from figures 15 and 16. The effective temperature, calculated from figure 18, is 17.2°C, close to that at Cedros Island, while the temperateness index is ten points lower, reflecting the wider annual range of temperature in the gulf. Apparently the high level of climatic warmth (effective temperature) enables the Surian-Panamic fauna to live in the northern part of the gulf, in spite of lower winter temperatures and a less equable climate.

The annual range of monthly mean temperatures at Santa Barbara, at the latitude of Elsmere Canyon, is from 12° to 18°C (Hall, 1964: fig. 2, table 11; my figs. 15, 16). The effective temperature, calculated from figure 18, is 14.5°C. The mean annual temperature is approximately 16°C and the mean annual range of temperature, approximately 9°C (table 8). The temperateness index, from figure 19, is between 70 and 71.

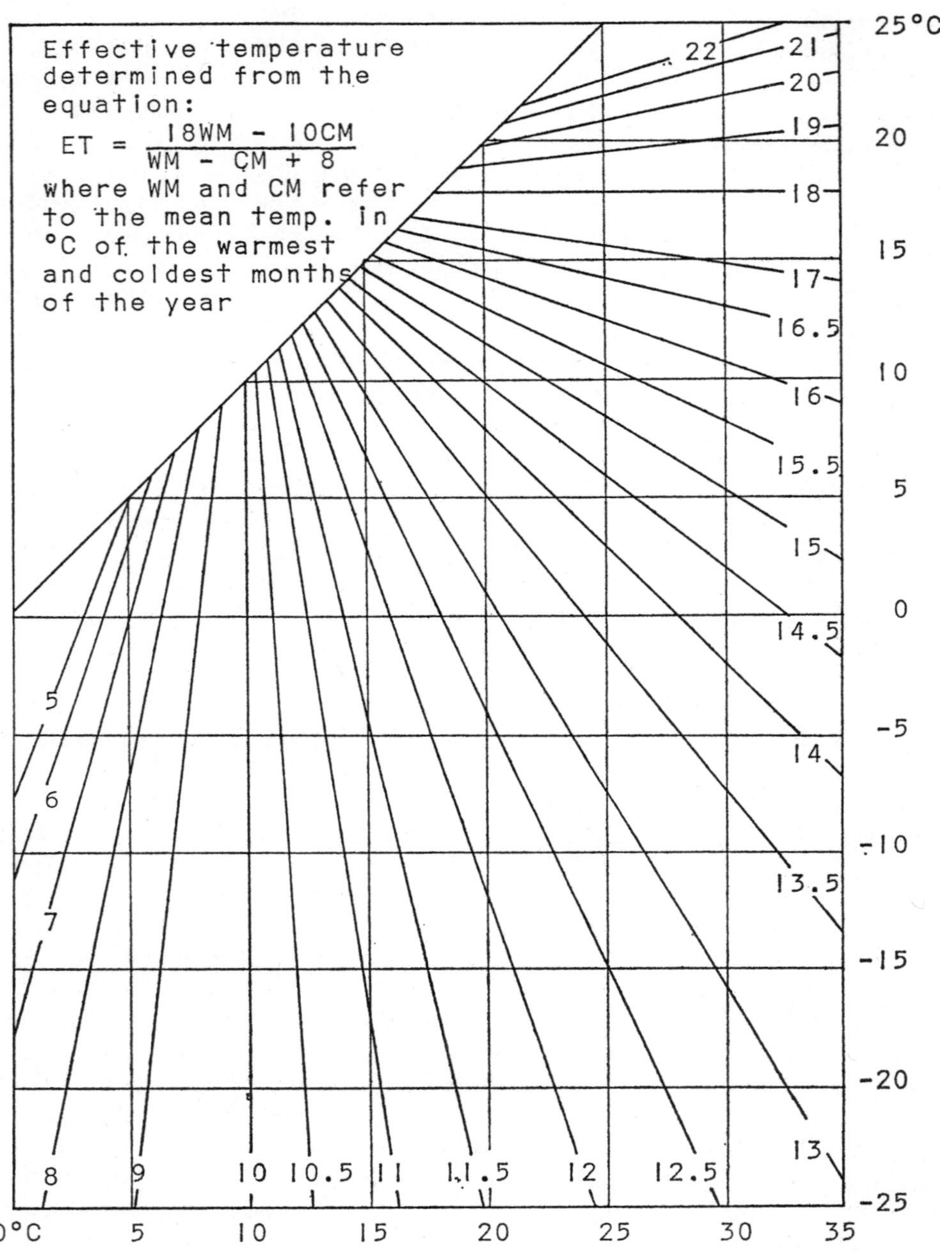

Fig. 18. Nomogram for determination of effective temperature. Modified from Bailey (1960).

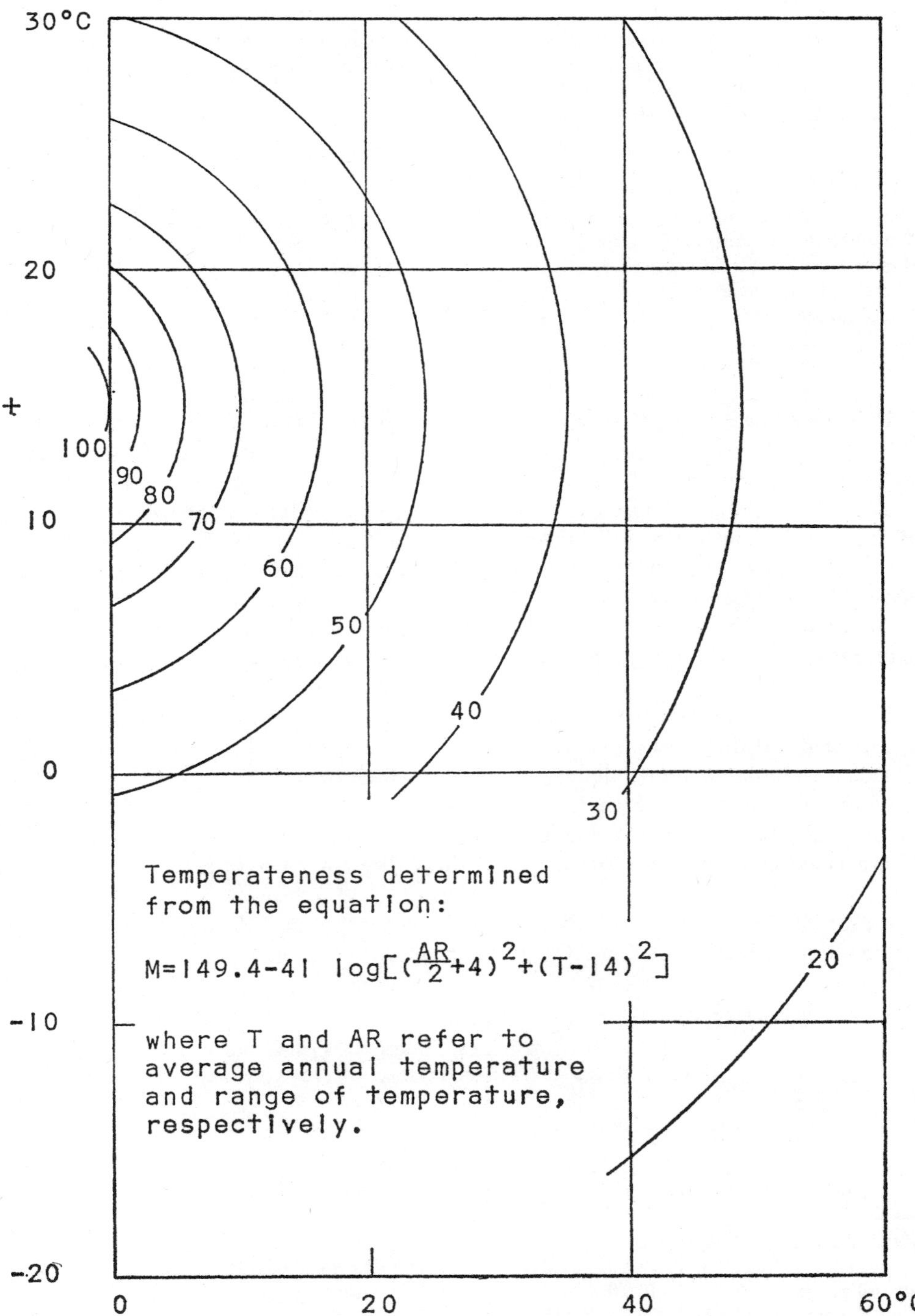

Fig. 19. Nomogram for determination of temperateness. Modified from Bailey (1964).

TABLE 7

Temperature Data for Surface Marine Water at Santa Barbara, Cedros Island, and North End of Gulf of California

Temperature data	Santa Barbara	Cedros Island	North end of Gulf of California
Mean annual temperature (°C)	±16	19–20	21–22
Annual range of monthly mean temperature (°C)	12–18	17–22	16–30
Mean annual range of temperature (°C)	±9	7–8	±13
Effective temperature (°C)	14.5	17.4	17.2
Temperateness index	70–71	68	±58

Note: Sources of data are given in text.

It is suggested above that the Surian-Panamic molluscan fauna is able to live at the north end of the Gulf of California, in spite of cold winters, because of the long warm summers. While the present-day marine climate at Santa Barbara is as equable as that at Cedros Island and more equable than in the northern end of the gulf, climatic warmth is several degrees lower than at those places. Mean monthly winter temperatures are also lower, but winter cold apparently is not so critical in restricting distribution of these taxa as is lack of summer warmth. The presence of the Surian-Panamic element at this latitude in early Pliocene time may reflect somewhat higher summer temperatures and slightly higher annual climatic warmth.

I have referred above to the role of the California Current in moving cold water southward and displacing summer isotherms to the south along the coast of California. A similar summer deflection of cold water toward the equator occurs on the west coast of South America, Africa, and to a lesser extent Australia (Sverdrup et al., 1942: charts 2, 3), indicating that this phenomenon is not unusual. There is no evidence that similar conditions did not exist during late Tertiary time.

Ingle (1967: 278–291; see discussion above) attributed Tertiary marine climatic changes in the southern California region to changes in polar climate and

TABLE 8

Mean Annual Surface-Water Marine Temperature and Annual Range of Temperature at Santa Barbara (In °C)

Temperature data	Year							
	1956	1957	1958	1962	1963	1964	1965	1966
Maximum	19.70	20.90	20.50	19.5	21.80	20.20	20.20	19.90
Minimum	11.10	10.60	12.60	11.0	11.80	11.10	11.70	12.00
Annual range	8.60	10.30	7.90	8.5	10.00	9.10	8.50	7.90
Annual mean	14.95	15.74	16.54	15.2	16.37	15.56	15.81	15.74

Note: Data from Scripps Institution of Oceanography (1960, 1965–1967).

consequently in the temperature of water in the California Current. This distribution of this cold south-flowing water off southern California is determined in part by the configuration of the coast, and climatic changes in this region also may reflect changes in coastal geography.

Figure 1 shows that early Pliocene geography of central and southern California was quite different from that of today. The interpretation shown in that figure is based on those of Reed (1933*a*), Reed and Hollister (1936), Corey (1954), and Axelrod (1956), and on my own analysis of available geologic maps of the region. Although the map is, of course, somewhat generalized and approximate, the seaways certainly existed over at least as large an area as shown, and the distribution of the major land masses is probably accurate. Movement on the San Andreas fault since early Pliocene time has displaced the basin south and east of Coalinga between 30 and 65 miles to the south relative to the area west of the fault (Crowell, 1962:18–20).

Clearly the effects of a south-flowing current of cold water would be quite different on this early Pliocene coastline. While the current may have cooled the outer coast of the large island north of Point Conception, this land area must have shielded the extensive coastal basins from its influence. Summer isotherms probably were normally distributed in these basins and more closely spaced only at about 36° north latitude, where the Salinas Valley seaway opened to the ocean and was exposed to the influence of the coastal current. Probably the range end points of species that are restricted to the north by summer breeding temperature requirements were concentrated here in a narrow latitudinal zone, as at Point Conception today. Under these conditions many molluscan taxa might have lived several degrees of latitude north of their Recent range end points throughout the area of the sheltered coastal basins. Only north of about 36° north latitude would summer temperatures have been depressed by the effects of the cold current from the north. Southern taxa would not be expected in early Pliocene faunas north of that latitude, and it is shown above that they are indeed rare or absent there.

Solar heating of water in the sheltered, deeply embayed coastal basins of southern California also may have been in part responsible for higher summer temperatures. This phenomenon would have had much the same effects as the inferred changes in the California Current, raising summer temperatures in southern and central California but not north of the Salinas Valley seaway.

Further evidence is provided by the early Pliocene distribution of Elsmere Canyon species that have Recent range end points near the latitude of Santa Barbara. The Recent ranges of eight of these species, from figure 13, are compared in figure 20 with their early Pliocene ranges. Of the eight species, some of which may be prevented from living farther north by the abruptly colder summer temperatures near Point Conception, four lived farther north in early Pliocene time. All of these had northern range end points close to 36° north latitude.

Figures 15 and 16 show that a reduction of the influence of the cold California Current could easily account for an increase in summer temperatures of as much as 4 degrees and in winter temperatures of about 1 degree. Effective temperature and temperateness for these suggested changes are presented in table 9

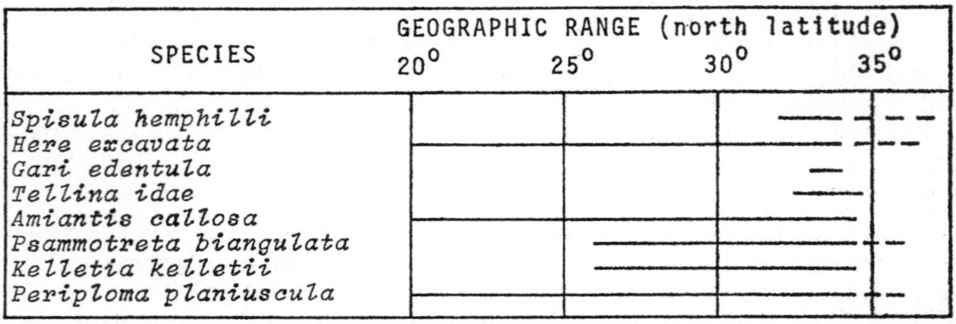

SPECIES — GEOGRAPHIC RANGE (north latitude) 20° 25° 30° 35°

Spisula hemphilli
Here excavata
Gari edentula
Tellina idae
Amiantis callosa
Psammotreta biangulata
Kelletia kelletii
Periploma planiuscula

Fig. 20. Recent and early Pliocene geographic ranges of southern species in the Elsmere Canyon fauna. Recent ranges are shown by solid lines; early Pliocene northward extensions of ranges are shown by dashed lines.

for comparison with the data in table 7. Temperateness, which probably is not so critical a factor in the sea as on the land, where survival temperatures are more limiting than temperatures required for reproduction, is not very different from today. Effective temperature, which reflects annual warmth or summer warmth and apparently is critical in determining molluscan distribution, is intermediate between those of Santa Barbara and the Surian molluscan province, with a value about halfway between those two extremes. This finding is consistent with the difference in faunal composition. The Surian-Panamic element in the Elsmere Canyon fauna is relatively small; the dominant element is that of the Californian province, which includes the latitude of Elsmere Canyon. The difference of about 1 degree in effective temperature possibly would be sufficient to allow a small number of taxa from the Surian and Panamic provinces to extend their ranges into this region. That nearly all the species that did so are now extinct may be owing to their relatively short geographic ranges and their evolution in isolated basins. When reduced summer temperatures made the Ventura basin region uninhabitable, they may not have been able to migrate to the south because of competition from established species in those regions.

CONCLUSIONS

The geographic ranges of all but a few of the Recent species in the Elsmere Canyon fauna overlap today about 1 degree of latitude south of Elsmere Canyon,

TABLE 9

Recent and Inferred Early Pliocene Effective Temperature and Temperateness at Santa Barbara

Temperature data	Recent	Early Pliocene
Mean annual temperature (°C)	16	17
Annual range of monthly mean temperature (°C)	12–18	13–22
Mean annual range of temperature (°C)	9	11
Effective temperature (°C)	14.5	15.7
Temperateness index	70–71	66

suggesting an early Pliocene thermal regime slightly warmer than that at the same latitude today. The few species that do not live at this latitude today, together with a number of extinct species in genera that also do not live at this latitude today, form two thermally anomalous elements in the fauna.

Six species in the Elsmere Canyon fauna, four living and two extinct, suggest water colder than today in early Pliocene time. As noted above, at least two of these taxa—*Mya* and *Patinopecten*—are of doubtful thermal significance. All six taxa probably either have evolved physiologically since early Pliocene time or are not limited geographically by temperature.

The following fourteen taxa are listed in table 3 as possible representatives of the Surian and Panamic molluscan provinces: *Anadara trilineata, Arca terminumbonis, Chione (Anomalocardia) fernandoensis, Dosinia jacalitosana, Miltha xantusi, Lyropecten estrellanus, Cymatium elsmerense, Astele* aff. *A. rema, Cantharus* sp., *Calicantharus fortis angulatus, C. humerosus, Ficus (Trophosycon) ocoyana, Bulla* cf. *B. punctulata*, and *Olivella* aff. *O. gracilis*. The thermal implications of *Anadara trilineata, Arca terminumbonis, Dosinia jacalitosana*, and *Miltha xantusi* are uncertain, as suggested above. The genus *Lyropecten* today does not live north of Scammon's Lagoon; *Cymatium* does not live north of San Ignacio Lagoon; *Cantharus* does not live north of Magdalena Bay; and *Astele* does not live north of the Tres Marías Islands. These clearly are thermally anomalous taxa. The extinct genus *Calicantharus* apparently is closely related to *Cantharus*. The genus *Ficus* does not live today north of Cape San Lucas, though the extinct subgenus *Trophosycon* is morphologically quite distinct from living subgenera. Uncertainly identified taxa that may have Surian-Panamic affinities include *Bulla* cf. *B. punctulata* and *Olivella* aff. *O. gracilis*. The former lives south of Guaymas today and the latter, south of Magdalena Bay (Keen, 1958:496, 424). This small group of taxa suggests that the marine thermal regime at Elsmere Canyon in early Pliocene time was different from the thermal regime in that region today.

It is suggested that early Pliocene marine climate in this region differed from present-day climate in having slightly higher summer temperatures, similar winter temperatures, and only slightly higher annual warmth. This interpretation may be consistent both with the overlapping ranges of Recent species in the Elsmere Canyon fauna and with the thermally anomalous southern element in the fauna. The differences from present-day climate may have been a result of the deflection of the south-flowing cold-water California Current away from the deeply embayed coastal basins by a large land mass in the area of the present coast between Point Conception and Monterey. Early Pliocene thermal conditions would have been different from today as far north as Monterey, north of which the effects of the California Current must have remained unchanged. Solar heating of large protected coastal basins in the southern and central California regions may also have had a role in raising summer temperatures. This interpretation is supported by the thermal implications of lower Pliocene fossil faunas along the entire coast. It is inferred that because of the higher summer temperatures and longer warm season east and south of the coastal barrier, some molluscan taxa were able to live farther north than at present in the Ventura basin and the Salinas-Coalinga region. The Salinas Valley fauna seems to have a warmer-water aspect than that of the Coalinga area, possibly because the latter was then between 30 and 65 miles north of its present position relative to rocks west of the San Andreas fault. The early Pliocene fauna of the Santa Cruz Mountains and the Eel River basin apparently were in the area of influence of the cold northern water and contained few southern molluscan taxa. These conclusions are summarized in figures 21 through 26, which show my interpretation of early Pliocene compared with Recent distribution of summer and winter isotherms and effective temperature isotherms along the California coast.

COMPARISON OF PRESENT INTERPRETATION WITH THOSE OF PREVIOUS STUDIES

Whereas I believe that early Pliocene marine climate in the Ventura basin was characterized by slightly higher summer and yearly warmth, previous studies held that winter temperatures were substantially higher in early Pliocene time than today. This conclusion apparently was based in most studies on the assumption that the geographic ranges of marine molluscs are limited by survival temperatures, while my interpretation takes into account other possible limiting factors. The median of midpoints-of-ranges index is consistent with my interpretation. Nearly all previous interpretations of the thermal significance of lower Pliocene molluscan faunas in the Coalinga region, the Salinas Valley, the Santa Cruz Mountains, and the Eel River basin also have emphasized southern extralimital taxa and have suggested substantially warmer early Pliocene climate. Planktonic foraminiferal evidence for substantially higher early Pliocene temperatures throughout California is also opposed to my interpretation. Early Pliocene marine climate in New Zealand apparently was similar to present-day climate at the same latitude, further suggesting that faunal differences in southern California were produced by local environmental factors, such as the configuration of the coastline. The general warmth of terrestrial climate in the southern California region also was similar to that existing there today, though apparently there was a narrower annual range of temperature. The different configuration of the coastline, however, with consequent effects on the course of the California Current and on solar heating of sheltered embayments, may account for the postulated broader range of temperature in the sea. With the exception of foraminiferal distribution, my interpretation seems to be consistent with most other kinds of evidence for early Pliocene shallow-water marine climate in the northeastern Pacific Ocean.

GEOLOGIC HISTORY

The Eocene basin of deposition must have extended east of the area of Elsmere and Grapevine canyons because several thousand feet of marine sediments were deposited here. Uplift of a pre–San Gabriel range of mountains took place after, and possibly in part during, this period of deposition, resulting in vertical elevation of the rocks east of the Whitney Canyon fault by at least 1,000 feet. By the time of deposition of the upper Miocene Modelo Formation the basin margin had been eroded down to a low, level surface. It was onto this gently sloping shore

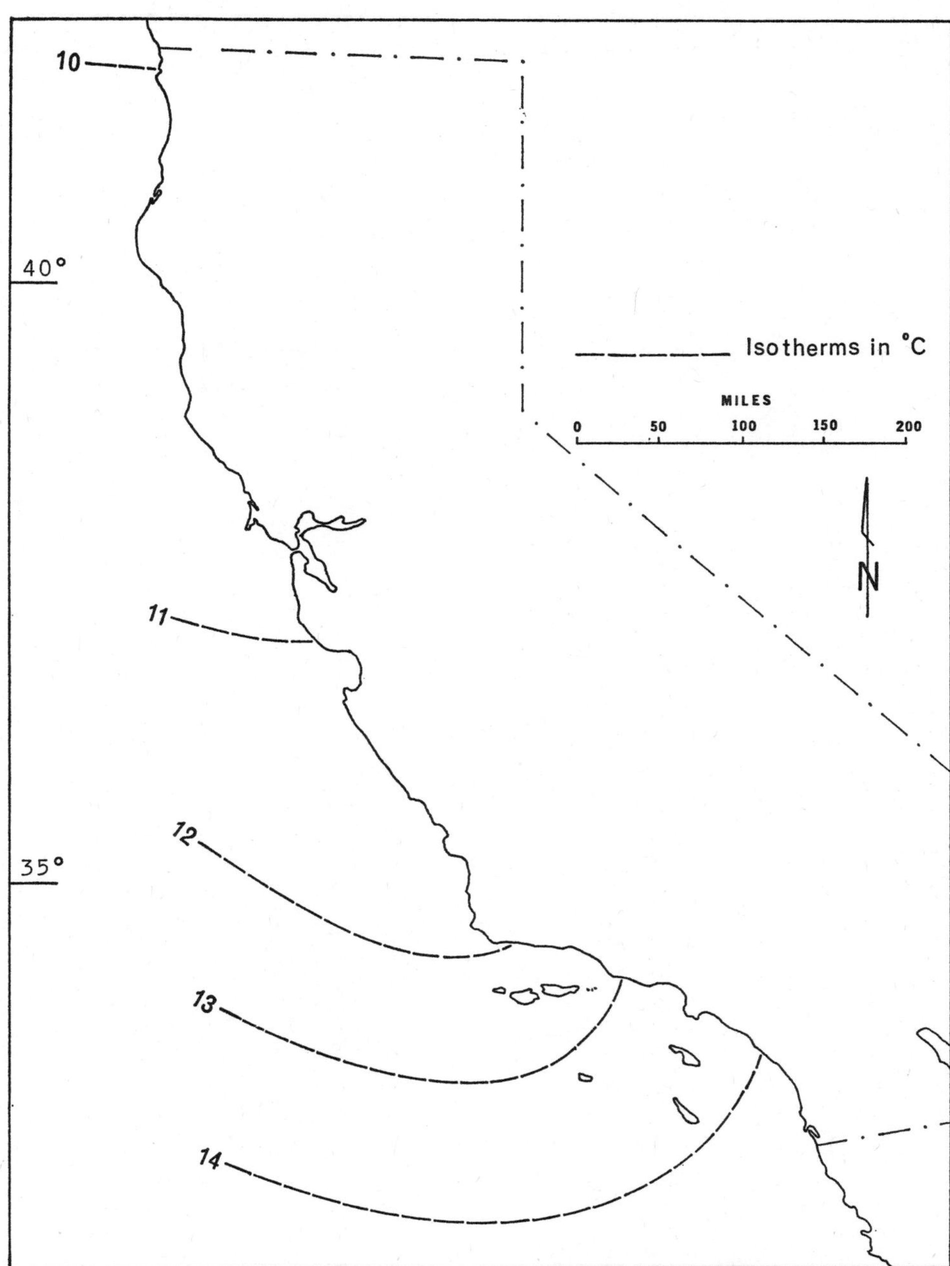

Fig. 21. Recent minimum winter isotherms. Data from figure 16.

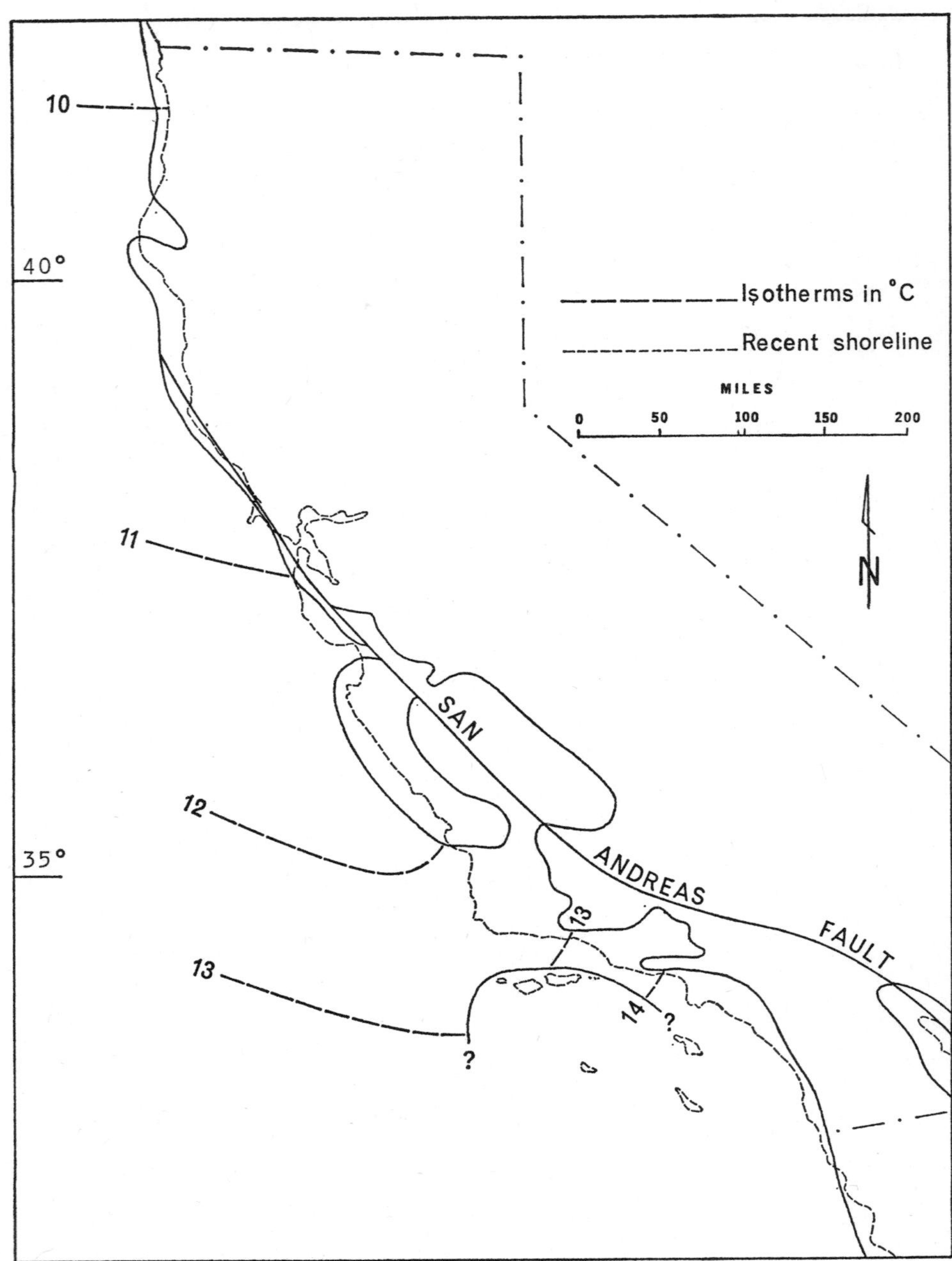

Fig. 22. Inferred positions of early Pliocene minimum winter isotherms. The marine basin east of the San Andreas fault is restored to its probable early Pliocene position relative to west side of fault.

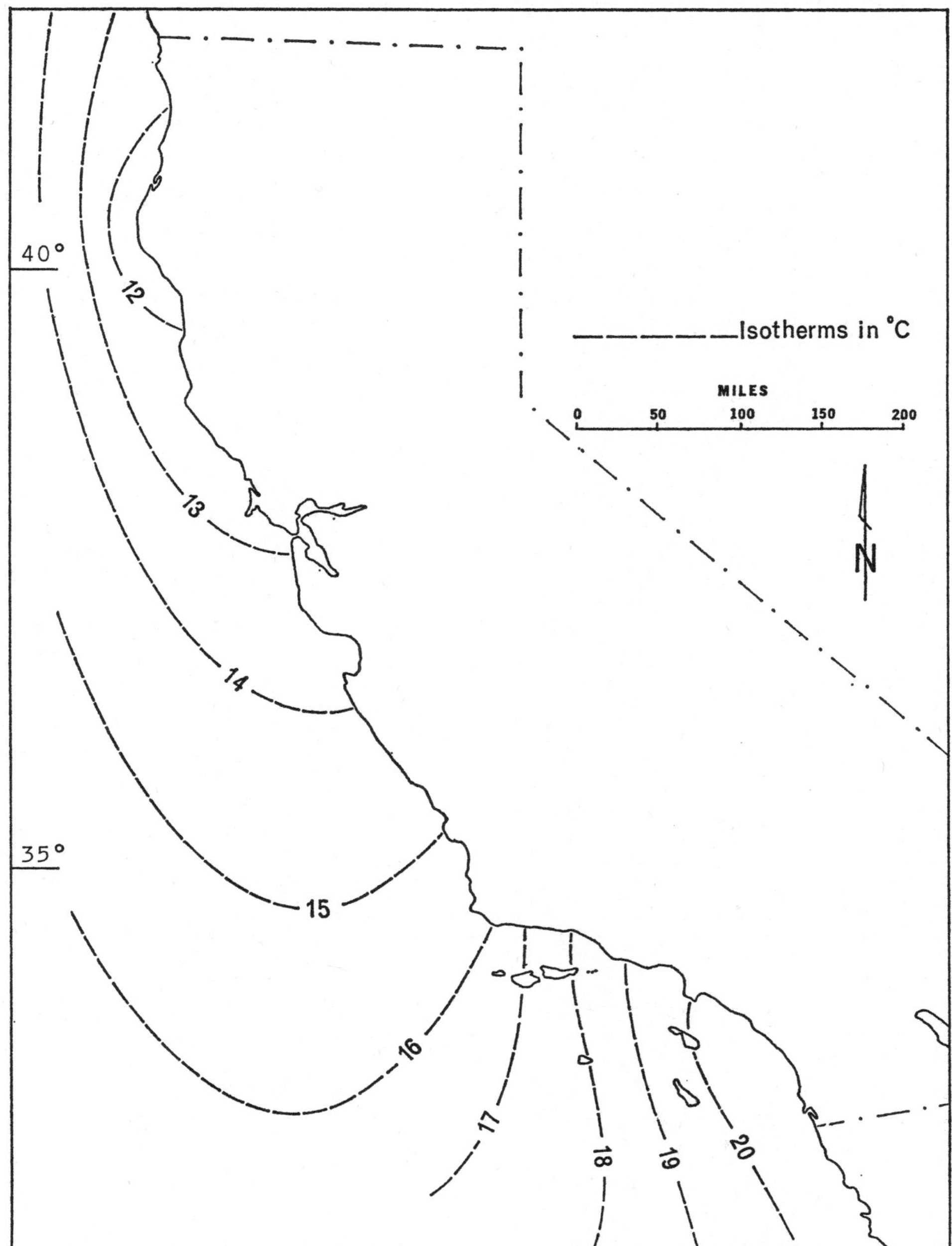

Fig. 23. Recent maximum summer isotherms. Data from figure 15.

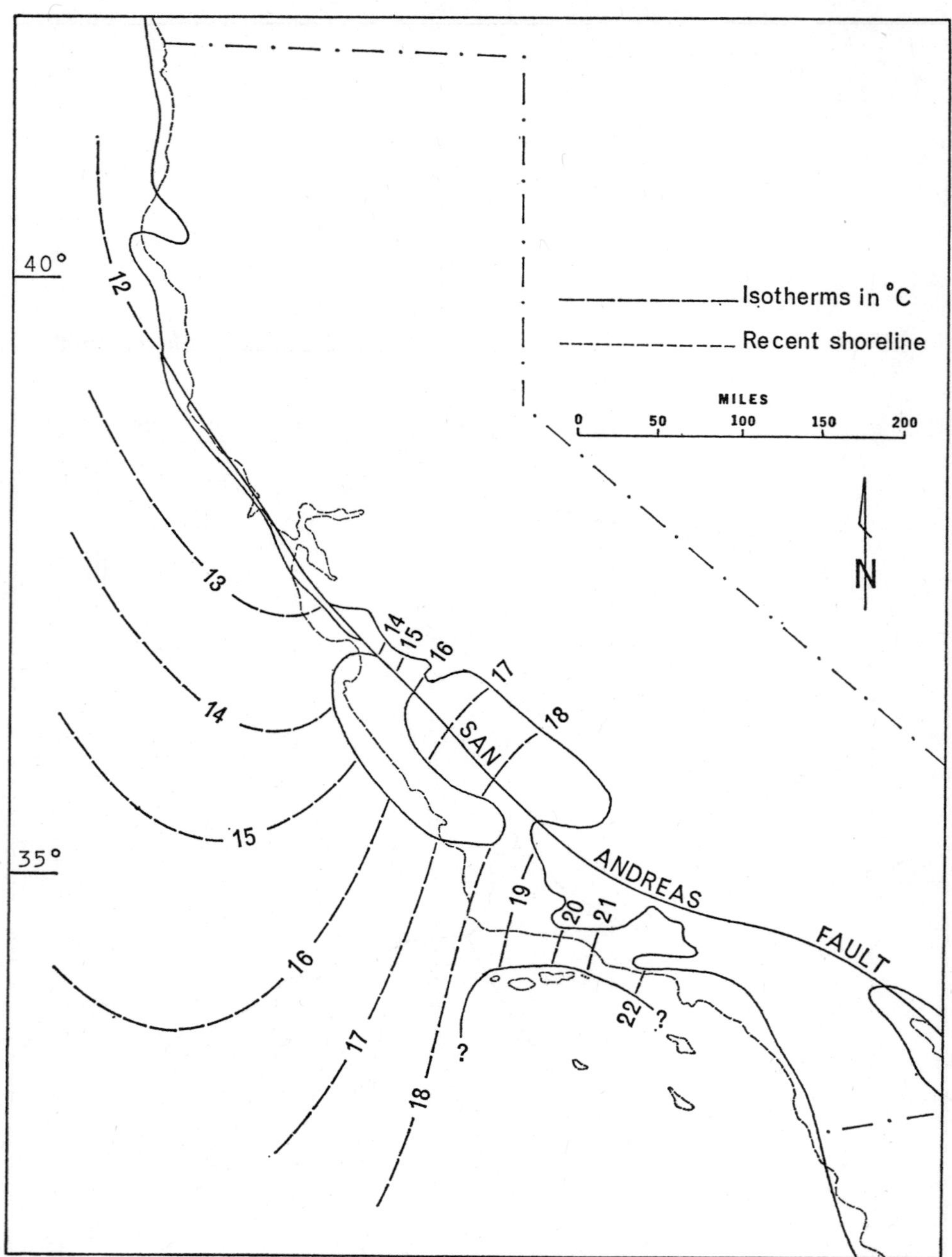

Fig. 24. Inferred positions of early Pliocene maximum summer isotherms. The marine basin east of the San Andreas fault is restored to its probable early Pliocene position relative to west side of fault.

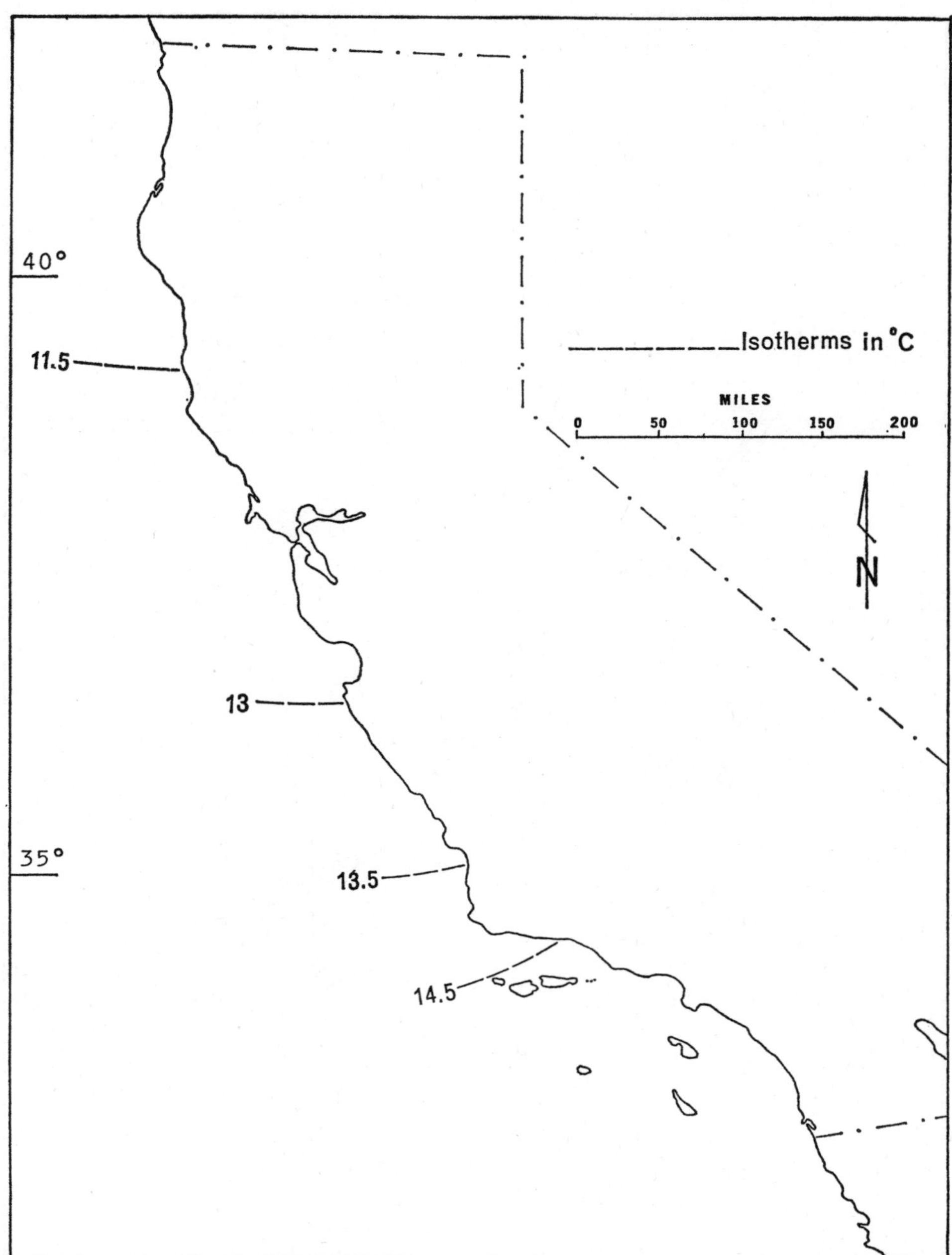

Fig. 25. Recent isotherms of effective temperature, based on data in figures 21 and 23.

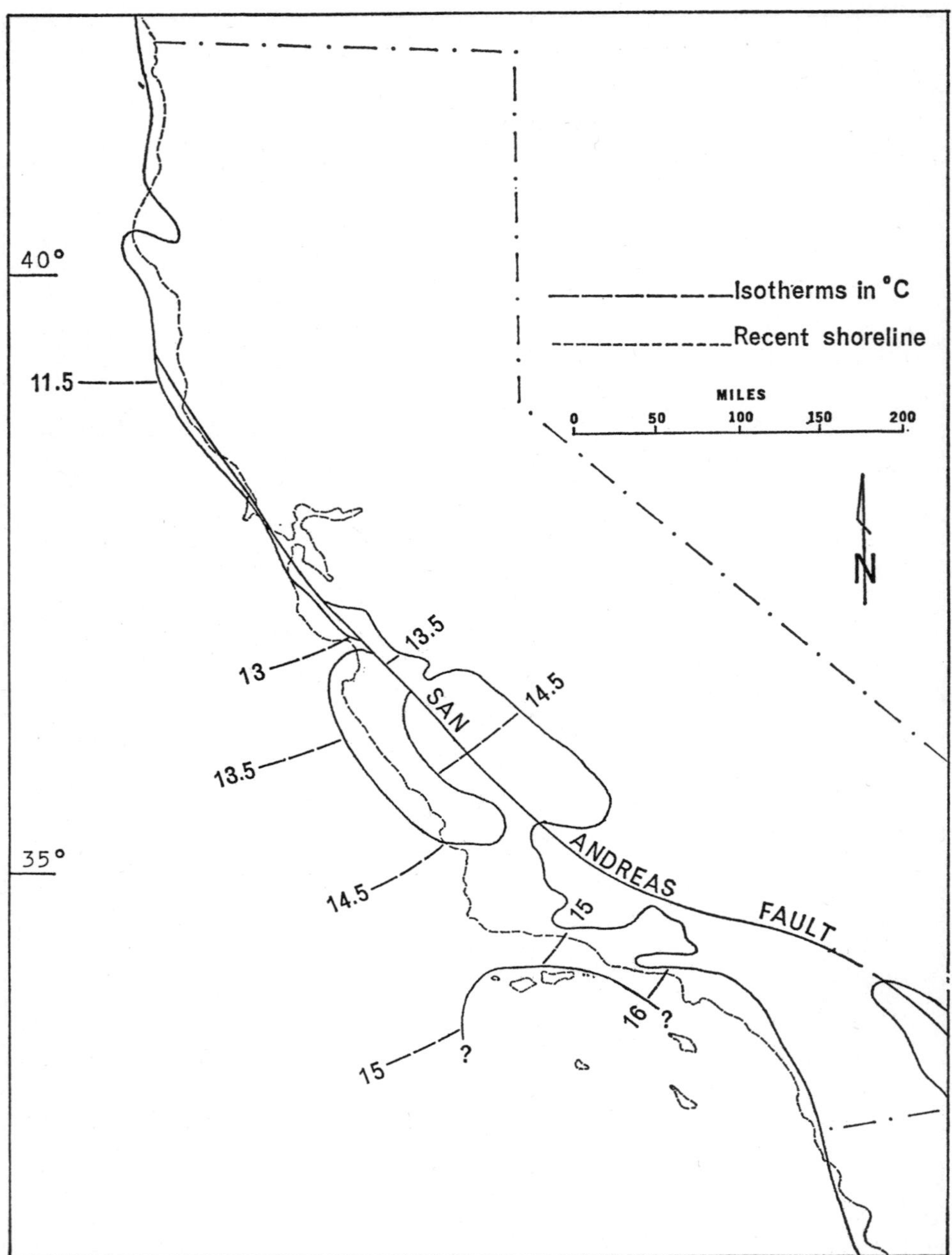

Fig. 26. Inferred positions of early Pliocene isotherms of effective temperature, based on data in figures 22 and 24. The marine basin east of the San Andreas fault is restored to its probable early Pliocene position relative to west side of fault.

that the late Miocene–early Pliocene sea transgressed, as the upward arching Santa Monica Mountains separated the Ventura and Los Angeles basins.

The Towsley Formation represents the final phase of a period of continuous marine deposition extending from early Miocene time to the middle of the Pliocene Epoch. The Elsmere Canyon rocks were deposited in shallow water just offshore at the eastern end of the Ventura basin. These shallow, nearshore waters supported a molluscan fauna similar to that at the same latitude today. Also present were barnacles, bryozoans, echinoderms, sharks and whales, and probably many of the soft-bodied, unpreserved faunal elements that live today in the nearshore marine environment. The coastal topography was probably also similar to that of present-day coastal California, with deep marine basins adjacent to young, rugged mountains. The vegetation on those mountains was little different from that growing there today, and apparently fires were common then as now. Generally speaking, the early Pliocene landscape of this region and its fauna and flora had a very modern aspect.

After a minor break in deposition and slight erosion of the youngest Towsley deposits, marine sediments of the Pico Formation filled the eastern end of the shrinking basin, grading upward into nonmarine deposits of the Saugus Formation (Winterer and Durham, 1962:317). Uplift of the present San Gabriel Range occurred during late Pliocene or early Pleistocene time, apparently after the region had been reduced to a low, level plain. The principal uplift at this time took place along the Grapevine Canyon fault and may have been accompanied by minor readjustments along the planes of older faults. A series of terraces on the north flank of the range, from 50 to 500 feet above the present Santa Clara River, reflects the intermittent nature of this deformation.

SYSTEMATIC PALEONTOLOGY

Because this study is primarily paleoenvironmental, "vertical" rather than "horizontal" relationships are emphasized in the following classification. That is, emphasis is placed on presumed genetic relationships between fossil and living species rather than on fine distinctions between stratigraphically superposed species. Further, because most of the species in this fauna are well known and have been described in available literature, systematic material in this section is minimal. Few subspecies are distinguished and subgeneric names are omitted except when useful or necessary for identification or to show relationships to living species. The brief synonymies include only those references necessary to show the origin of the names used in this paper and to record the names used in all published lists of the Elsmere Canyon fauna. Complete references to works cited in synonymies are included in the Literature Cited at the end of the paper.

Because of their varied usage, the qualifying symbols "aff." and "cf." are defined as follows: "aff." indicates that the specimens are morphologically similar and probably genetically related to, but in some characters distinct from, the species with which they are compared; "cf." indicates that the specimens are incomplete or poorly preserved, and features critical to specific identification are lacking or unobservable; a high degree of probability of identity is implied.

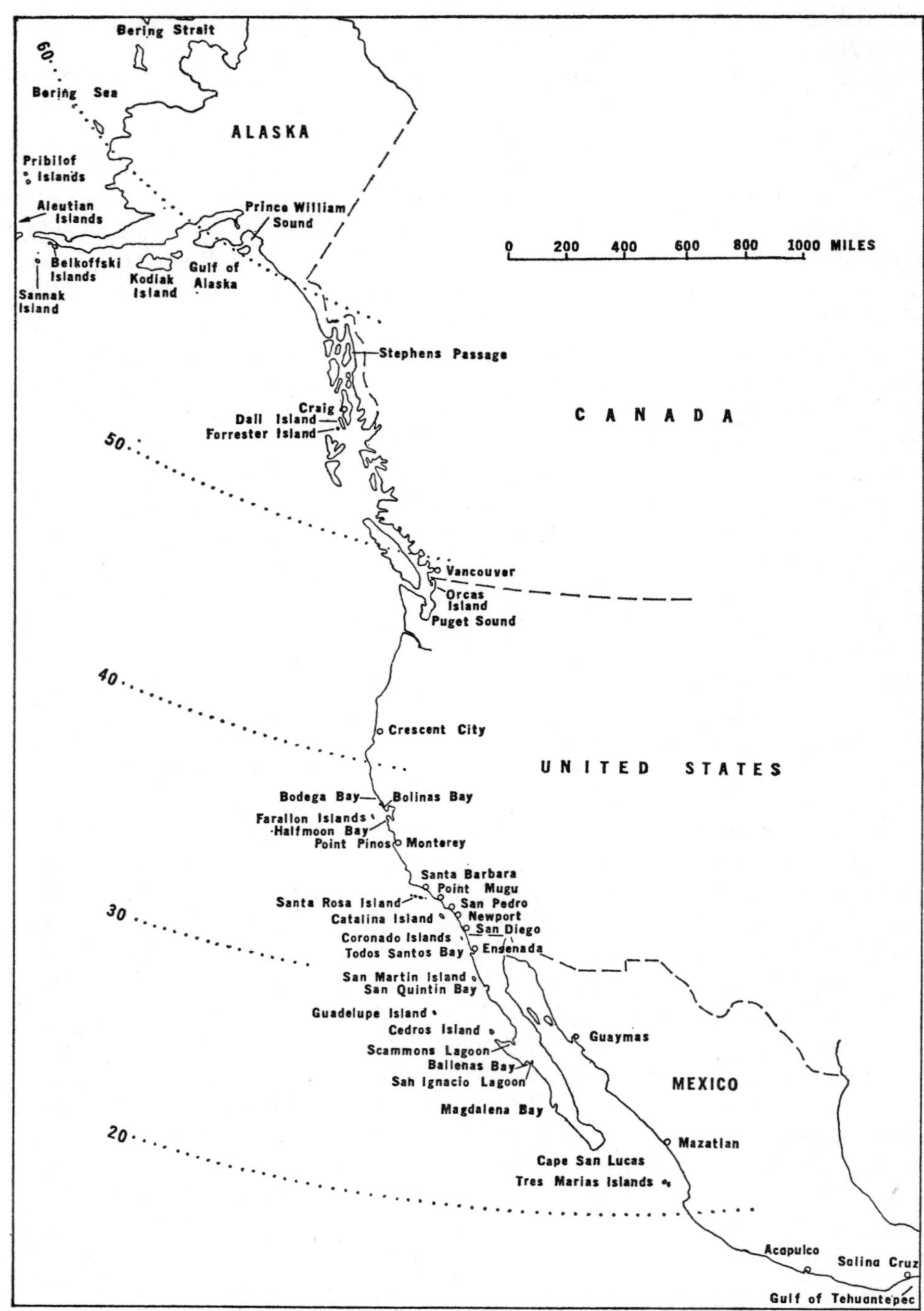

Fig. 27. Map of the northeast Pacific coast with localities referred to in Systematic Paleontology section.

TABLE 10

SYMBOLS USED IN SYSTEMATIC PALEONTOLOGY SECTION TO REFER TO SOURCES OF DATA ON AGE AND ENVIRONMENT OF SPECIES

Symbol	Source
Add	Addicott, 1965
Ade	Adegoke, 1969
Ba	Bandy, 1958
B	Burch, 1944–1946
D	Dall, 1921
GG	Grant and Gale, 1931
HS	Hertlein and Strong, 1940–1951
J	Jordan, 1924
K	Keen, 1958, 1963
L	Light et al., 1964
M	McLean, 1969
P	Packard, 1918
RC	Ricketts and Calvin, 1939
S	Stanton, 1966
SG	Smith and Gordon, 1948
W	Woodring, 1938

Ecological data are included in the discussions of all living species in the fauna. Under "Distribution" are given Recent ranges of living species. Range end-point localities are shown on the map in figure 27. Depth ranges are given in fathoms. Sources of data on age and ecology are abbreviated as shown in table 10.

Phylum BRYOZOA
Class GYMNOLAEMATA

Remarks.—Unidentified encrusting bryozoans were collected at five localities.

Phylum MOLLUSCA
Class BIVALVIA
Order NUCULOIDA
Family NUCULIDAE
Genus *Acila* H. and A. Adams
Acila castrensis (Hinds, 1843)

Nucula castrensis Hinds, 1843*a*:98; English, 1914: 210, 213; Kew, 1924: 78.
Acila castrensis Hinds. Cooper, 1888:227.
Nucula (Acila) castrensis Hinds. Grant and Gale, 1931:116–117, pl. 1, figs. 6*a*, 6*b*.

Hypotype.—UCLA no. 48008, locality 2055.
Age.—Miocene to Recent (GG:116).
Distribution.—Sitka to Cedros Island (B: 33:8).
Depth.—15 to 700 fms, abundant below 40 fms (Ba:707; B:33:8; P:247; SG:168; W:13).
Substrate.—Fine sand and mud (B:33:8; P:247; SG:168).

Family Nuculanidae
Genus *Nuculana* Link
Nuculana taphria (Dall, 1896)

Leda taphria Dall, 1896:70, new name for *L. caelata* Hinds; English, 1914:210; Kew, 1924:78.
Nuculana taphria (Dall). Grant and Gale, 1931:121, pl. 1, figs. 8, 9.

Hypotype.—UCLA no. 48009, locality 5782.
Age.—Late Miocene to Recent (Ade:89).
Distribution.—Bodega Bay to Arena Bank (K, 1958:19).
Depth.—6 to 80 fms (K, 1958:19; Ba:706; B:33:11; SG:169; P:248).
Substrate.—Fine sand and silt (B: 33:11; SG:169; P:248).

Order ARCOIDA
Family Arcidae
Genus *Arca* Linné
Arca terminumbonis Grant and Gale, 1931

Arca (*Navicula*) *terminumbonis* Grant and Gale, 1931:142–143, pl. 1, figs. 18*a–c*, 19*a–c*, 20*a–c*, 22.

Hypotype.—UCLA no. 48010, locality 5799.
Age.—Early Pliocene of Elsmere Canyon (GG:142).

Remarks.—This distinctive species is known only from Elsmere Canyon. It occurs in the basal beds in at least three localities and may have been byssal on rocks in shallow water.

Genus *Anadara* Gray
Anadara trilineata (Conrad, 1856)

Arca trilineata Conrad, 1856:314; Arnold, 1907*a*:526, 544, pl. 48, figs. 3, 3*a*, 4; Eldridge and Arnold, 1907:25, 107, pl. 38, figs. 3, 4; English, 1914:209; Smith, 1919:152; Kew, 1924:77.
Arca (Arca) trilineata Conrad. Grant and Gale, 1931:139–140, pl. 2, figs. 1, 4.
Anadara trilineata (Conrad). Schenck and Keen, 1940: pl. 50, figs. 1, 2.

Hypotype.—UCLA no. 48011, locality 2056.
Age.—Late Miocene to late Pliocene (GG:140; Ade:91).

Remarks.—This species is abundant at several localities, but specimens are smaller than normal everywhere except at UCLA locality L-2066 in the upper unit in lower Elsmere Canyon.

Genus *Barbatia* Gray
Barbatia cf. *B. bailyi* (Bartsch, 1931)

Acar bailyi Bartsch, 1931:2, pl. 1, 5 central figs.
Barbatia (*Acar*) *bailyi* (Bartsch). Keen, 1958:26.

Age.—Recent (K, 1958:26).
Distribution.—Santa Monica to Panama (M:67).

Remarks.—The single specimen has been lost from the collection, so there is no hypotype number.

Barbatia pseudoillota Reinhart, 1937

Barbatia (*Fugleria*) *pseudoillota* Reinhart, 1937:184–185, pl. 28, figs. 6, 9, 10.

Hypotype.—UCLA no. 48012, locality 2056.
Age.—Early and middle Pliocene (Reinhart, 1937:185).

Remarks.—This species was not collected during the present study, but a single specimen is in the UCLA collections (UCLA locality L-2056).

Family GLYCYMERIDAE

Genus *Glycymeris* da Costa

Glycymeris profunda (Dall, 1879)

Axinea profunda Dall, 1879:13–14.
Glycymeris profunda Dall. Willet, 1943:111–113, pl. 11, figs. 3, 3*a*.

Hypotype.—UCLA no. 48013, locality 5769.
Age.—Early Pliocene (this report); early Pleistocene to Recent (Willet, 1943:112–113).
Distribution.—Redondo Beach and Catalina Island (Willet, 1943:112–113; B:34:3).
Depth.—25 to 200 fms (Willet, 1943:112–113).

Order MYTILOIDA

Family MYTILIDAE

Genus *Lithophaga* Röding

Lithophaga cf. *L. plumula* (Hanley, 1844)

Lithodomus plumula Hanley, 1844:17.
Lithophagus plumula Hanley. Carpenter, 1857 [1855–1857]: 125–126.
Lithophaga plumula Hanley. Dall, 1898:799.
Lithophaga plumula (Hanley). McLean, 1969:68, fig. 37.4.

Hypotype.—UCLA no. 48014, locality 5765.
Age.—Pleistocene to Recent (GG: 253).
Distribution.—Mendocino County to Peru (M:68).
Depth.—Intertidal to 35 fms (SG: 171).
Substrate.—Boring in shells or limestone (K, 1963:106).

Remarks.—The single specimen collected is not encrusted. Otherwise it is identical in size and form with *L. plumula*.

Genus *Modiolus* Lamarck

Modiolus rectus (Conrad, 1837)

Modiola recta Conrad, 1837:243, pl. 19, fig. 1.
Modiolus rectus Conrad. Dall, 1898:793; Arnold, 1907*a*:526; Eldridge and Arnold, 1907:25.
Volsella recta (Conrad). Grant and Gale, 1931: 249–250.

Hypotype.—UCLA no. 48015, locality 2056.
Age.—Oligocene(?), Miocene to Recent (GG:250).
Distribution.—Vancouver Island (M:67) to Conception Bay, Gulf of Calif. (K, 1958:56).
Depth.—Intertidal to 20 fms (SG:171; B:36:13).
Substrate.—Byssal in mud, solitary (SG:171).

Genus *Mytilus* Linné

Mytilus coalingensis Arnold, 1909

Mytilus (Mytiloconcha) coalingensis Arnold, 1909:73–75, pl. 19, fig. 5; pl. 22, fig. 6.

Hypotype.—UCLA no. 48016, locality 5800.
Age.—*Pliocene* (Ade: 95).

Order PTERIOIDA

Family PECTINIDAE

Genus *Lyropecten* Conrad

Lyropecten estrellanus (Conrad, 1856)

Pallium estrellanum Conrad, 1856:313.
Pecten estrellanus Conrad var. *catalinae* Arnold, 1906:76, pl. 20, figs. 3, 3*a*, 4; Arnold, 1907*a*: 527; Eldridge and Arnold, 1907:25.

Pecten ashleyi Arnold. English, 1914:210; Smith, 1919:152; Kew, 1924:78.
Pecten cerrosensis Gabb. English, 1914:210; Smith, 1919:152; Kew, 1924:78.
Pecten (*Lyropecten*) *estrellanus* (Conrad) variety *catalinae* Arnold. Grant and Gale, 1931:186.
Pecten (*Lyropecten*) *estrellanus* (Conrad) variety *cerrosensis* Gabb. Grant and Gale, 1931: 187, pl. 8, figs. 1*a*, 1*b*, 2*a*, 2*b*; pl. 9, fig. 2.
Lyropecten cerrosensis (Gabb). Woodring, 1938:32–35, pl. 7, figs. 1, 2.

Hypotype.—UCLA no. 48017, locality 5800.
Age.—Late Miocene to early Pliocene (GG: 185–187; Ade: 101).

Genus *Patinopecten* Dall

Patinopecten lohri (Hertlein, 1928)

Pecten (*Patinopecten*) *oweni* Arnold, 1906:63, pl. 8, figs. 1, 1*a*, 1*b*. Not *Pecten oweni* De Gregorio, 1884.
Pecten (*Patinopecten*) *lohri* Hertlein, 1928:93–94.
Pecten healeyi Arnold, 1907*a*:527; Eldridge and Arnold, 1907:25; Smith, 1919:152; Kew, 1924:78.
Pecten healyi Arnold. English, 1914:210, 213.
Pecten oweni Arnold. English, 1914:210, 213; Smith, 1919:152; Kew, 1924:78: not De Gregorio, 1884.
Pecten (*Patinopecten*) *healeyi* Arnold variety *lohri* (Hertlein). Grant and Gale, 1931:197–198, pl. 6, figs. 1*a*, 1*b*.
Patinopecten lohri (Hertlein). Woodring, Stewart, and Richards, 1940:70, 91–92, pl. 35, figs. 2–5.

Hypotype.—UCLA no. 48018, locality 5803–2.
Age.—Early to middle Pliocene (GG:197–198; Ade: 103).

Family ANOMIIDAE

Genus *Pododesmus* Philippi

Pododesmus cepio (Gray, 1850)

Placunanomia cepio Gray, 1850:121.
Pododesmus (*Monia*) *macroschisma* Deshayes. Dall, 1898:780–781; not *Anomia macroschisma* Deshayes, 1839.
Monio machroschisma Deshayes. Arnold, 1907*a*:527.
Monia machroschisma Dall. Eldridge and Arnold, 1907:25.
Pododesmus macroschisma (Deshayes). Grant and Gale, 1931:241–242, pl. 12, figs. 3, 4*a*, 4*b*.

Hypotype.—UCLA no. 48019, locality 2056.
Age.—Early Pliocene to Recent (GG:242).
Distribution.—British Columbia to Gulf of Calif. (K, 1958:80).
Depth.—Intertidal to 35 fms (B: 36:3).
Substrate.—On rocks and shells (B:36:3).

Remarks.—The earlier named *P. macroschisma* Deshayes is restricted to extreme north Pacific waters (K, 1958:80).

Family LIMIDAE

Genus *Lima* Bruguiere

Lima hemphilli Hertlein and Strong, 1946

Lima (*Limaria*) *hemphilli* Hertlein and Strong, 1946 [1940–1951]:66–67, pl. 1, figs. 3, 4.

Hypotype.—UCLA no. 48020, locality 2056.
Age.—Pliocene to Recent (GG:239, as *L. dehiscens*).
Distribution.—Monterey to Acapulco (K, 1958:75).
Habitat.—Free-swimming (K, 1958:75).

Order VENEROIDA
Family Lucinidae
Genus *Here* Gabb
Here excavata (Carpenter, 1857)

Lucina excavata Carpenter, 1857 [1855–1857]:98.
Phacoides richthofeni Gabb. English, 1914:210; Kew, 1924:78.
Lucina (*Here*) *excavata* Carpenter. Stewart, 1930:181–183, pl. 15, fig. 3; pl. 17, fig. 5; Grant and Gale, 1931:290, pl. 14, figs. 2, 5, 10.

Hypotype.—UCLA no. 48021, locality 5770.
Age.—Oligocene to Recent (GG:290–291).
Distribution.—San Pedro to Mazatlan (K, 1958:94).
Depth.—Intertidal to 66 fms (B:40:6; K, 1958:94).

Genus *Lucina* Bruguiere
Lucina californica Conrad, 1837

Lucina californica Conrad, 1837:255, pl. 20, fig. 1.

Hypotype.—UCLA no. 48022, locality 2056.
Age.—Pliocene to Recent (GG: 286).
Distribution.—Crescent City to San Ignacio Lagoon (D:35).
Depth.—Intertidal to 40 fms (B:40:7).

Remarks.—This species was not collected during the present study, but several specimens are in the UCLA collections.

Genus *Lucinisca* Dall
Lucinisca nuttallii (Conrad, 1837)

Lucina nuttallii Conrad, 1837:255, pl. 20, fig. 2.
Phacoides (*Lucinisca*) *nuttallii* Conrad. Dall, 1901*b*:812.
Phacoides nuttali Conrad. English, 1914:210; Kew, 1924:78.
Lucina (*Myrtea*) *nuttallii* Conrad. Grant and Gale, 1931:288, pl. 14, figs. 4*a*, 4*b*, 18.

Hypotype.—UCLA no. 48023, locality 5790.
Age.—Early Miocene to Recent (GG:288; Ade:114).
Distribution.—Monterey (M:74) to Tres Marías Islands (K, 1958:96).
Depth.—Intertidal to 25 fms (B:40:7; SG:173).
Substrate.—Sand and mud (B:40:7; SG:173; K, 1958:96).

Genus *Lucinoma* Dall
Lucinoma annulata (Reeve, 1850)

Lucina annulata Reeve, 1850 [1843–1878]: v. 6, sp. 17, pl. 4, fig. 17.
Phacoides annulatus Reeve. Arnold, 1907*a*:527; Kew, 1924:78.
Phacoides acutilineatus Conrad. Eldridge and Arnold, 1907:25; English, 1914:210, 213.
Lucina (*Myrtea*) *acutilineata* Conrad. Grant and Gale, 1931:286–287, pl. 14, figs. 22*a*, 22*b*.
Lucina (*Lucinoma*) *annulata* Reeve. Olsson, 1961:209, pl. 30, figs. 3, 3*a*, 3*b*.

Hypotype.—UCLA no. 48024, locality 5773.
Age.—Late Miocene to Recent (Ade:113).
Distribution.—Alaska (D:35) to Cedros Island and Gulf of Calif. (HS).
Depth.—4 to more than 400 fms (Ba:706; P:264; B:40:7); deeper than 30 fms in Lower Calif. (K, 1958:97).

Genus *Miltha* H. and A. Adams
Miltha xantusi (Dall, 1905)

Phacoides (*Miltha*) *xantusi* Dall, 1905:111.

Phacoides santaecrucis Arnold. English, 1914:210.
Phacoides sanctaecrucis Arnold. Smith, 1919:152; Kew, 1924:78.
Lucina (*Miltha*) *xantusi* (Dall). Grant and Gale, 1931:291–292, pl. 14, figs. 20*a*, 20*b*.

Hypotype.—UCLA no. 48025, locality 5771.
Age.—Miocene to Recent (GG: 291).
Distribution.—Cape San Lucas (K, 1958:98).

Remarks.—This species apparently was abundant in shallow water during early Pliocene time in southern California (see text).

Family Ungulinidae
Genus *Diplodonta* Bronn
?Diplodonta sericata (Reeve, 1850)

Lucina sericata Reeve, 1850 [1843–1878]: v. 6, sp. 55, pl. 9, fig. 55.
?Diplodonta serricata Reeve. Carpenter, 1857:248.

Hypotype.—UCLA no. 48026, locality 5794.
Age.—Pleistocene to Recent (GG: 295).
Distribution.—Rare at Monterey; common from San Ignacio Lagoon to Ecuador (K, 1958: 103).
Depth.—4 to 40 fms (K, 1958:103).

Remarks.—A single valve from UCLA locality 5794 probably is referable to this species, but the hinge is not preserved.

Family Carditidae
Genus *Cyclocardia* Conrad
Cyclocardia californica (Dall, 1903)

Venericardia (*Cyclocardia*) *californica* Dall, 1903: 1431–1432, pl. 56, fig. 16.
Venericardia californica Dall. English, 1914:210; Kew, 1924:78.
Cardita ventricosa Gould. Grant and Gale, 1931:272 (in part).
Cyclocardia californica (Dall). Woodring, Bramlette, and Kew, 1946:82.

Hypotype.—UCLA no. 48027, locality 5805.
Age.—Miocene to Recent (GG: 273).
Distribution.—Belkofski Bay, Alaska, to Coronado Islands (as *Venericardia ventricosa*—D:32).

Remarks.—Because of the similarity of this species to others living farther south, it has not been used in the interpretation of climate.

Family Cardiidae
Genus *Nemocardium* Meek
Nemocardium centifilosum (Carpenter, 1864)

Cardium var. *centifilosum* Carpenter, 1864:611, 642.
Protocardia centifilosa Carpenter. English, 1914:210.
Laevicardium (*Nemocardium*) *centifilosum* (Carpenter). Grant and Gale, 1931:311, pl. 19, figs. 9, 10.

Hypotype.—UCLA no. 48028, locality 5766.
Age.—Miocene(?), Pliocene to Recent (GG:311).
Distribution.—Bodega Bay to Ensenada (B:41:27)
Depth.—10 to 46 fms (SG:174; P:268; B:41:27).

Genus *Trachycardium* Morch

Trachycardium quadragenarium (Conrad, 1837)

Cardium quadragenarium Conrad, 1837:230, pl. 17, fig. 5.
Cardium (*Trachycardium*) *quadragenarium* Conrad. Dall, 1901*a*:389.
Cardium quadrigenarium Conrad var. *fernandoensis* Arnold, 1907*a*:526, 535, pl. 68, figs. 2, 2*a*; Eldridge and Arnold, 1907:25, pl. 38, fig. 2; English, 1914:209; Kew, 1924:77.
Cardium quadrigenarium Conrad. Smith, 1919:152.
Laevicardium (*Trachycardium*) *quadragenarium* (Conrad) variety *fernandoense* (Arnold). Grant and Gale, 1931:307.
Trachycardium (*Dallocardia*) quadragenarium (Conrad). Woodring, 1938:53, pl. 9, fig. 1.

Hypotype.—UCLA no. 48029, locality 5780.
Age.—Miocene to Recent (GG:306).
Distribution.—Monterey (SG: 174) to southern Baja California (M:76).
Substrate.—Sand (SG:174; K, 1963:107).

Family Mactridae

Genus *Mactra* Linné

Mactra cf. *M. dolabriformis* (Conrad, 1867)

Spisula dolabriformis Conrad, 1867:193.
Mactra dolabriformis Conrad. Dall, 1894*a*:138, pl. 5, fig. 1.

Hypotype.—UCLA no. 48030, locality 5785.
Age.—Pleistocene to Recent (GG:393).
Distribution.—Lobitas, Calif. (D:51), to Panama (K, 1958:156).

Remarks.—These specimens resemble *M. californica* Conrad and *M. falcata* Gould but are most similar to *M. dolabriformis,* differing only in being slightly more elongate.

Genus *Spisula* Gray

Spisula hemphilli (Dall, 1894)

Mactra hemphilli Dall, 1894*a*:137–138, pl. 5, fig. 2.
Spisula (*Hemimactra*) *hemphilli* Dall, 1894*b*:40.
Mactra cf. *hemphilli* Dall. Arnold, 1907*a*:526; Eldridge and Arnold, 1907:25.
Mactra (*Spisula*) *hemphilli* Dall. Grant and Gale, 1931:398.

Hypotype.—UCLA no. 48031, locality 5788.
Age.—Early Pliocene to Recent (GG:398; Ade:122).
Distribution.—Santa Barbara to Ensenada (M:82).
Depth.—Intertidal to 25 fms (B:44:20).
Habitat.—Sand of exposed beaches (B:44:20).

Genus *Tresus* Gray

Tresus nuttallii (Conrad, 1837)

Lutraria (*Cryptodon*) *nuttallii* Conrad, 1837:235, pl. 18, fig. 1.
Tresus nuttallii Conrad. Dunker, 1882, Index Moll. Maris Japonici, p. 184 [according to Grant and Gale, 1931:404].

Hypotype.—UCLA no. 48032, locality 2056.
Age.—Late Miocene to Recent (GG: 405; Ade:123).
Distribution.—Prince William Sound, Alaska, to Scammon's Lagoon (GG:405).

Remarks.—This species was not collected during the present study, but a single specimen is in the UCLA collections (UCLA locality L-2056).

Family SOLENIDAE
Genus *Solen* Linné
Solen sicarius Gould, 1850

Solen sicarius Gould, 1850*b*:214; English, 1914:209, 210, 214; Kew, 1924:78; Grant and Gale, 1931:385–386, pl. 21, fig. 4.

Hypotype.—UCLA no. 48033, locality 5786.
Age.—Late Miocene to Recent (GG:385; Ade:134).
Distribution.—Vancouver Island to San Quintin Bay (D:50).
Depth.—Intertidal to 68 fms (SG:176; P:281; K, 1963:107).
Substrate.—Sand and mud (K, 1963:107; B:43:27).

Family TELLINIDAE
Genus *Macoma* Leach
Macoma indentata Carpenter, 1864

Macoma indentata Carpenter, 1864:611, 639; Arnold, 1903:161–162, pl. 16, fig. 1; 1907*a*:526; Eldridge and Arnold, 1907:25; English, 1914:210; Kew, 1924:78; Grant and Gale, 1931: 374–375.

Hypotype.—UCLA no. 48034, locality 2055.
Age.—Late Miocene to Recent (GG:374–375; Ade:129).
Distribution.—Puget Sound (GG:375) to Scammon's Lagoon (J:150).
Depth.—9 to 30 fms (SG:175; P:278; Ba:706).

Macoma nasuta (Conrad, 1837)

Tellina nasuta Conrad, 1837:258–259.
Macoma nasuta Conrad. Carpenter, 1864:639; English, 1914:210.
Macoma nasuta (Conrad). Grant and Gale, 1931:365–366, pl. 20, figs. 11*a*, 11*b*.

Hypotype.—UCLA no. 48035, locality 5786.
Age.—Oligocene(?), Miocene to Recent (GG:366).
Distribution.—Kodiak Island to Scammon's Lagoon (D:47).
Depth.—Intertidal to 25 fms (B:43:11–12).

Macoma secta (Conrad, 1837)

Tellina secta Conrad, 1837:257–258.
Macoma secta Conrad. H. and A. Adams, 1858:401.
Macoma secta (Conrad). Grant and Gale, 1931:374, pl. 20, figs. 6*a*, 6*b*.

Hypotype.—UCLA no. 48036, locality 5803-3.
Age.—Late Miocene to Recent (GG:374; Ade:128).
Distribution.—Vancouver (D:48) to southern Baja California (M:84).
Depth.—Intertidal to 25 fms (B:43:16).

Macoma yoldiformis Carpenter, 1864

Macoma yoldiformis Carpenter, 1864:602, 611, 639; Oldroyd, 1924:177, pl. 44, fig. 6.

Age.—Early Pliocene (this report); middle Pliocene to Recent (GG:373).
Distribution.—Dall Island, Alaska, to San Diego (B:43:15).
Depth.—Intertidal to 25 fms (B:43:15).

Remarks.—The single specimen has been lost so there is no hypotype.

Genus *Psammotreta* Dall
Psammotreta (*Florimetis*) *biangulata* (Carpenter, 1856)

?*Scrobicularia biangulata* Carpenter, 1856:230.

Metis alta Conrad. English, 1914:210; Smith, 1919:152; Kew, 1924:78.
Apolymetis biangulata (Carpenter). Grant and Gale, 1931:363–364, pl. 20, fig. 16.
Psammotreta (*Florimetis*) *biangulata* (Carpenter). Keen, 1962:178.

Hypotype.—UCLA no. 48037, locality 5778.
Age.—Miocene to Recent (GG:364).
Distribution.—Point Conception to Magdalena Bay (M:85).
Depth.—Intertidal to 25 fms (B:43:9).

Genus *Tellina* Linné
Tellina idae Dall, 1891

Tellina idae Dall, 1891:183–185, pl. 6, fig. 3; pl. 7, figs. 1, 4; Arnold, 1907*a*:527; Eldridge and Arnold, 1907:25; English, 1914:210; Kew, 1924:78; Grant and Gale, 1931:358, pl. 20, figs. 12, 14*a*, 14*b*.

Hypotype.—UCLA no. 48038, locality 5775.
Age.—Miocene to Recent (GG:358).
Distribution.—Santa Barbara to San Diego (M:83).
Depth.—Shallow to 50 fms (Valentine, 1961: table 11).

Family Psammobiidae
Genus *Gari* Schumacher
Gari californica (Conrad, 1849)

Psammobia californica Conrad, 1849:121; Oldroyd, 1924:185, pl. 43, fig. 5.
Gari californica (Conrad). Grant and Gale, 1931:382.

Hypotype.—UCLA no. 48039, locality 2066.
Age.—Early Pliocene to Recent (GG:382).
Distribution.—Alaska to Magdalena Bay (M:86).
Depth.—Intertidal to 25 fms, more common offshore (B:43:21).
Substrate.—Sand (B:43:21).

Gari edentula (Gabb, 1869)

?Siliquaria edentula Gabb, 1869 [1866–1869]:53, pl. 15, fig. 11
Gari edentula (Gabb). Grant and Gale, 1931:382–383, pl. 21, fig. 5

Hypotype.—UCLA no. 48040, locality 5775.
Age.—Middle Miocene to Recent (GG:383; Ade:124).
Distribution.—San Pedro to San Diego (D:49).
Depth.—Common at 15 fms (B:43:22).

Family Veneridae
Genus *Amiantis* Carpenter
Amiantis callosa (Conrad, 1837)

Cytherea callosa Conrad, 1837:252.
Amiantis callosa Conrad. Carpenter, 1864:526, 640; Arnold, 1907*a*:526, 544, pl. 49, fig. 2; Eldridge and Arnold, 1907:25, pl. 39, fig. 2.
Amiantis cf. *callosa* Carpenter. English, 1914:209; Smith, 1919:152.
Amiantis cf. *A. callosa* Carpenter. Kew, 1924:77.
Amiantis callosa (Conrad). Grant and Gale, 1931:348–349, pl. 17, figs. 7, 9, 11–14.

Hypotype.—UCLA no. 48041, locality 5782.
Age.—Late Miocene to Recent (GG:348; Ade:137).
Distribution.—Santa Barbara (Conrad, 1837:252) to Gulf of Tehuantepec, Mexico (GG:348).
Depth.—Intertidal to 10 fms (K, 1963:104; B:42:7).
Substrate.—Sand of exposed beaches (B:42:7).

Genus *Securella* Parker

Securella elsmerensis English, 1914

Chione elsmerensis English, 1914:209, 214, pl. 23, figs. 1*a*, 1*b*; Smith, 1919:152; Kew, 1924:78.
Venus (*Chione*) *elsmerensis* (English). Grant and Gale, 1931:319, pl. 16, figs. 6*a*, 6*b*, 7.
Securella elsmerensis (English). Parker, 1949:590, pl. 93, figs. 14, 16; pl. 94, fig. 9.

Hypotype.—UCLA no. 48042, locality 5801.
Age.—Late Miocene (S:23) to early Pliocene (GG:319; Parker, 1949:590).

Remarks.—Only a single partial specimen, including the left hinge, was collected.

Genus *Chione* Megerle

Chione (*Anomalocardia*) *fernandoensis* English, 1914

Chione n. sp. Arnold, 1907*a*:526.
Chione fernandoensis English, 1914:209, 215, pl. 23, figs. 9*a*, 9*b*; Kew, 1924:78.
Venus (*Chione*) *securis* Shumard variety *fernandoensis* (English). Grant and Gale, 1931:321, pl. 17, figs. 4*a*, 4*b*, 5, 6.
Chione (*Anomalocardia*) *fernandoensis* English. Parker, 1949:585, pl. 95, figs. 7, 8, 13–16, 18.

Hypotype.—UCLA no. 48043, locality 2055.
Age.—Late Miocene (S:23) to early Pleistocene (Parker, 1949:585; Ade:139).

Genus *Compsomyax* Stewart

Compsomyax subdiaphana (Carpenter, 1864)

?*Clementia subdiaphana* Carpenter, 1864:602, 607, 640.
Callista subdiaphana Carpenter. Arnold, 1907*a*:526, 544, pl. 49, fig. 3; Eldridge and Arnold, 1907:25, pl. 39, fig. 3.
Marcia subdiaphana Carpenter. English, 1914:210; Smith, 1919:152; Kew, 1924:78.
Venerella (*Compsomyax*) *subdiaphana* (Carpenter). Stewart, 1930:224.
Clementia (*Compsomyax*) *subdiaphana* Carpenter. Grant and Gale, 1931:334, pl. 17, figs. 10*a*, 10*b*, 15.
Compsomyax subdiaphana Carpenter. Burch, 1946 [1944–1946]:no. 42, p. 11; no. 43, p. 34; no. 45, p. 15.

Hypotype.—UCLA no. 48044, locality 5777.
Age.—Miocene(?) to Recent (GG:334).
Distribution.—Sannak Islands, Alaska (Oldroyd, 1924:47), to Todos Santos Bay (B:42:11); Cedros Island (HS).
Depth.—5 to 46 fms (K, 1963:105; P:268; SG:174; B:42:11).

Genus *Dosinia* Scopoli

Dosinia jacalitosana Arnold, 1909

Dosinia jacalitosana Arnold, 1909:67, pl. 16, fig. 5.
Dosinia ponderosa Gray (n. var.?) English, 1914:210; Kew, 1924:78.
Dosinia ponderosa Gray. Smith, 1919:152.

Hypotypes.—UCLA no. 48045, locality 2066; no. 48046, locality 5777.
Age.—Early Pliocene (Ade:145).

Remarks.—This species is very similar to *Dosinia ponderosa* Gray, a living species found from Scammon's Lagoon to Peru (B:42:6; K, 1958:136).

Genus *Protothaca* Dall

Protothaca cf. *P. staminea* (Conrad, 1837)

Venus staminea Conrad, 1837:250–251, pl. 19, fig. 15.

Paphia (Protothaca) staminea Conrad. Dall, 1902:397–399, pl. 14, fig. 2.

Hypotype.—UCLA no. 48047, locality 5800.
Age.—Miocene to Recent (GG:330).
Distribution.—Aleutian Islands to southern Baja California (M:78).

Remarks.—The single valve collected is very similar to Recent specimens of *P. staminea.*

Genus *Callithaca* Dall

Callithaca tenerrima (Carpenter, 1864)

Tapes tenerrima Carpenter, 1864:641; Arnold, 1907*a*:527.
Paphia tenerrima Carpenter. Smith, 1919:152.
Venerupis (Callithaca) tenerrima (Carpenter). Grant and Gale, 1931:327–328, pl. 18, figs. 9*a*, 9*b*.

Hypotype.—UCLA no. 48049, locality 2056.
Age.—Early Pliocene to Recent (Ade:147).
Distribution. Puget Sound to southern Baja Calif. (M:80).

Remarks.—This species was not collected during the present study, but one specimen is in the UCLA invertebrate fossil collection.

Genus *Saxidomus* Conrad

Saxidomus nuttallii Conrad, 1837

Saxidomus nuttallii Conrad, 1837:249–250, pl. 19, fig. 12.

Hypotype.—UCLA no. 48048, locality 2056.
Age.—Middle Miocene to Recent (GG:342; Ade:141).
Distribution.—Humboldt Bay to San Martín Island, Lower Calif. (GG:342).
Depth.—A few to 25 fms (B:42:8–9).

Remarks.—This species was not collected during the present study, but two specimens are in the UCLA collection.

Genus *Tivela* Link

Tivela stultorum (Mawe, 1823)

Donax stultorum Mawe, 1823:37, 40, pl. 9, fig. 7.
Tivela (Pachydesma) stultorum Mawe, Dall, 1902:386.

Hypotype.—UCLA no. 48050, locality 5787.
Age.—Pliocene to Recent (GG:340).
Distribution.—Halfmoon Bay (GG:340) to Magdalena Bay (M:78).
Depth.—Intertidal to a few fms (B:42:11; K, 1963:107).
Habitat.—Sand of exposed beaches (B:42:11; K, 1963:107).

Order MYOIDA

Family MYIDAE

Genus *Cryptomya* Conrad

Cryptomya californica (Conrad, 1837)

Sphaenia californica Conrad, 1837:234, pl. 17, fig. 11.
Cryptomya californica Conrad, 1849:121; Arnold, 1907*a*:526; Eldridge and Arnold, 1907:25.
Cryptomya cf. *ovalis* Conrad. English, 1914:210.
Cryptomya cf. *C. ovalis* Conrad. Kew, 1924:78.
Cryptomya californica (Conrad). Grant and Gale, 1931:417, pl. 21, figs. 7, 8*a*, 8*b*, 11, 14*a*, 14*b*.

Hypotype.—UCLA no. 48051, locality 5786.
Age.—Late Miocene to Recent (GG:418; Ade:152).
Distribution.—Alaskan Gulf to southern Mexico (K, 1958:207).
Depth.—Intertidal to 50 fms. B:43:18, 44:26; K, 1963:105; SG:177).

Genus *Mya* Linné

Mya truncata Linné, 1758

Mya truncata Linné, 1758:670; Arnold, 1907*a*:527, pl. 50, fig. 1; Eldridge and Arnold, 1907:25; Smith, 1919:152.

Mya (*Mya*) *truncata* Linné. Grant and Gale, 1931:414–415.

Hypotype.—UCLA no. 48052, locality 2056.
Age.—Miocene to Recent (GG:415).
Distribution.—Circumboreal, south to Puget Sound (D:52).

Family Hiatellidae

Genus *Panopea* Menard de la Groye

Panopea generosa Gould, 1850

Panopea generosa Gould, 1850*b*:215; Arnold, 1907*a*:527; Eldridge and Arnold, 1907:25; Smith, 1919:152.

Panopaea generosa Gould. English, 1914:210, 213.

Panope generosa Gould. Kew, 1924:78.

Panope (*Panope*) *generosa* Gould. Grant and Gale, 1931:424, pl. 21, figs. 12*a*, 12*b*.

Hypotype.—UCLA no. 48053, locality 5777.
Age.—Miocene to Recent (GG:424–425).
Distribution.—Forrester Island, Alaska (B:44:29), to Scammon's Lagoon (J).
Depth.—Intertidal to 25 fms (B:44:30).
Substrate.—Sand or mud (K, 1963:106; B:44:30).

Family Teredinidae

Genus *Bankia* Gray

Bankia cf. *B. setacea* (Tryon, 1863)

Xylotrya setacea Tryon, 1863:144–145, pl. 1, figs. 2, 3.

Bankia (*Bankia*) *setacea* Tryon. Bartsch, 1922:7–8, pl. 4, fig. 5; pl. 30, fig. 3.

Hypotype.—UCLA no. 48054, locality 5815.
Age.—Recent.
Distribution. —Kodiak Island to San Diego (RC:325).
Substrate.—Burrows in wood.

Remarks.—Fossil wood riddled with burrows of this taxon occurs at a number of localities. A single shell fragment and numerous pallets are preserved in a specimen from UCLA locality 5815.

Order PHOLADOMYOIDA

Family Periplomatidae

Genus *Periploma* Schumacher

Periploma planiuscula Sowerby, 1834

Periploma planiuscula Sowerby, 1834*b*:87; Grant and Gale, 1931:255–256, pl. 13, figs. 1*a*, 1*b*.

Hypotype.—UCLA no. 48055, locality 5786.
Age.—Early Pliocene (this report); middle Pliocene to Recent (GG:255).
Distribution.—Point Conception to Peru (B:37:12).
Depth.—Low tide to a few fms (K, 1958:229; B:37:12).

Family Thraciidae

Genus *Thracia* Leach

Thracia trapezoides Conrad, 1849

Thracia trapezoides Conrad, 1849. U.S. Exploring Expedition, v. 10, p. 723, pl. 17, fig. 6*a*. [according to Grant and Gale, 1931:276].

Hypotype.—UCLA no. 48056, locality 5803-2.

Age.—Oligocene to Recent (GG:257).

Distribution.—Alaska to Redondo Beach (B:37:13).

Depth.—Intertidal to 80 fms (Ba; B:37:13; K, 1963:107).

Class GASTROPODA
Subclass PROSOBRANCHIA
Order ARCHAEOGASTROPODA
Family ACMAEIDAE
Genus *Acmaea* Eschscholtz
Acmaea cf. *A. digitalis* Eschscholtz, 1833

Acmaea digitalis Eschscholtz, 1833:20, pl. 23, figs. 7, 8.

Hypotype.—UCLA no. 48057, locality 5786.

Age.—Recent.

Distribution.—Aleutian Islands to San Diego (RC:17).

Depth.—Uppermost intertidal (RC:17).

Remarks.—The single specimen is not well preserved. The apex is more anterior and depressed than is usual in this species, but some Recent specimens (e.g., UCLA cat. no. 20761) approach this form.

Family HALIOTIDAE
Genus *Haliotis* Linné
Haliotis fulgens Philippi, 1845

Haliotis fulgens Philippi, 1845, Zeitschar. f. Malak., p. 150 [according to Grant and Gale, 1931:846].

Hypotype.—UCLA no. 48059, locality 5759 (*Haliotis* sp.).

Age.—Pliocene to Recent (GG:846).

Distribution.—Farallon Islands and Catalina Island to Gulf of Calif. (GG:846).

Substrate.—Rocks.

Remarks.—One unidentifiable poorly preserved specimen of *Haliotis* was collected during this study. This species was reported from Elsmere Canyon by Grant and Gale (GG:846).

Family FISSURELLIDAE
Genus *Diodora* Gray
Diodora murina (Arnold, 1903)

Fissurella (*Glyphis*) *murina* "Carpenter" Dall, in Orcutt, 1886:543 [*nomen nudum,* according to Palmer, 1958:120].

Fissuridea murina Arnold, 1903:339.

Diodora murina (Carpenter MS). Dall, 1921:185.

Hypotype.—UCLA no. 48059.

Age.—Middle Pliocene to Recent (GG:850).

Distribution.—Crescent City to Magdalena Bay (D:185).

Remarks.—This species was not collected during the present study, but one specimen is in the UCLA collection.

Family TROCHIDAE
Genus *Calliostoma* Swainson
Calliostoma aff. *C. gemmulatum* Carpenter, 1864

Calliostoma gemmulatum Carpenter, 1864:612, 653; Grant and Gale, 1931:835, pl. 32, fig. 21.

Hypotype.—UCLA no. 48060, locality 2055.
Age.—Pliocene to Recent (GG:835).
Distribution.—Morro Bay to Gulf of Calif. (B:60:46).
Substrate.—Rocks and algae (K, 1963:100).

Remarks.—Four partial specimens were collected. The sculpture of three to five granular spiral ribs is indistinguishable from that of *C. gemmulatum,* but apparently the whorls are flat, not concavely shouldered as in that species.

Calliostoma splendens Carpenter, 1864

Calliostoma splendens Carpenter, 1864:653; Oldroyd, 1927, pt. 3:182–183, pl. 98, fig. 1.

Hypotype.—UCLA no. 48061, locality 5794.
Age.—Early Pliocene (this report); Pleistocene to Recent (GG:837).
Distribution.—Monterey to Guadalupe Island (Strong and Hanna, according to B:58:3).
Depth.—20 to 50 fms (B:58:3).

Genus *Astele* Swainson

Astele aff. *A. rema* (Strong, Hanna, and Hertlein, 1933)

Calliostoma rema Strong, Hanna, and Hertlein, 1933:121–122, pl. 5, figs. 3, 4.
Astele rema (Strong, Hanna, and Hertlein). Keen, 1958:256.

Hypotype.—UCLA no. 48062, locality 5783.
Age.—Recent.
Distribution.—Tres Marías Islands to Mazatlan (K, 1958:256).

Remarks.—Illustrations in Strong, Hanna, and Hertlein (1933: pl. 5, figs. 3, 4) and in Keen (1958:256) are not clear, but this single specimen is very close to the original description of *A. rema.* It has a very low spire and the body whorl has two distinct angulations, one at the basal periphery and another at one-third of the distance from there to the suture. The deep, open umbilicus is bounded by a rounded carina which forms a toothlike projection at the base of the columella. The columella is curved and reflected slightly over the umbilicus. The inner surface is apparently nacreous. The type of *A. rema* is 17 mm wide and 11 mm high; this specimen is 17 mm wide and 13 mm high. The difference in height appears to be due to the extreme length of the columella in this specimen. Since it extends anteriorly somewhat more than in the type, the lowest part of the aperture is at the tooth on the umbilical carina at the base of the inner lip. This specimen also differs in having more numerous spiral ribs. *A. rema* has three ribs above the upper keel, one between the keels, and eight on the base; this specimen has approximately ten, five, and twelve, respectively, in those positions.

Genus *Margarites* Gray

Margarites pupillus (Gould, 1849)

Trochus pupillus Gould, 1849:91.
Margarites (*Pupillaria*) *pupilla* Gould. Dall, 1921:178, pl. 17, figs. 2, 10.

Hypotype.—UCLA no. 48063, locality 5758.
Age.—Early Pliocene (this report); late Pliocene to Recent (GG:840).
Distribution.—Bering Sea to San Pedro (D:178) to San Diego in deep water (B:58:6).

Genus *Tegula* Lesson

Tegula gallina (Forbes, 1852)

Trochus (*Monodonta*) *gallina* Forbes, 1852:271, pl. 11, figs. 8*a*, 8*b*.

Tegula (*Chlorostoma*) *gallina* Forbes. Dall, 1921:174.
Tegula (*Chlorostoma*) *gallina* (Forbes). Grant and Gale, 1931:827–828, pl. 32, figs. 30, 31.

Hypotype.—UCLA no. 3321, locality 418.
Age.—Early Pliocene to Recent (GG:828).
Distribution.—Santa Barbara County to Magdalena Bay (M:22).
Depth.—Intertidal to 15 fms (K, 1963:104).
Substrate.—Rocks and algae (K, 1963:104).

Remarks.—One poorly preserved specimen was collected during this study, and two others are in the UCLA collections (UCLA cat. no. 3321).

Family TURBINIDAE
Genue *Astraea* Röding
Astraea gradata Grant and Gale, 1931

Astraea (*Pomaulax*) *gradata* Grant and Gale, 1931:818–819, pl. 31, figs. 1*a*, 1*b*, 3*a*, 3*b*, 5, 8, 9.

Hypotypes.—UCLA no. 48064, locality 5779; no. 48065, locality 5782.
Age.—Late Miocene (S:31–32) to middle Pliocene (Winterer and Durham, 1962: table 5).

Genus *Tricolia* Risso
?*Tricolia* sp.

Hypotype.—UCLA no. 48066, locality 5786.

Remarks.—A single poorly preserved specimen appears to belong in this genus.

Order MESOGASTROPODA
Family EULIMIDAE
Genus *Balcis* Gray
Balcis rutila (Carpenter, 1864)

Eulima rutila Carpenter, 1864:613, 659.
Melanella rutila Carpenter. Oldroyd, 1927: pt. 2:75, pl. 46, figs. 2, 3, 6.
Balcis rutila (Carpenter). Burch, 1945 [1944–1946]: no. 53:6, 8, 10.

Hypotype.—UCLA no. 48067, locality 5758.
Age.—Early Pliocene (this report); Pleistocene to Recent (GG:863).
Distribution.—Craig, Alaska, to Magdalena Bay (GG:863).
Depth.—Shallow to 100 fms (B:53:11).
Substrate.—Commensal on echinoderms (B:53:11).

Genus *Opalia* H. and A. Adams
Opalia varicostata Stearns, 1875

Opalia varicostata Stearns, 1875:463–464, pl. 27, figs. 2–5.

Hypotype—UCLA no. 48069, locality 5803-3.
Age.—Early Pliocene (see remarks) to middle Pliocene (GG:854).

Remarks.—I have collected this species from the lower Pliocene Capistrano Formation south of San Juan Capistrano, California.

Family LITTORINIDAE
Genus *Littorina* Ferussac
Littorina sp.

Hypotype.—UCLA no. 48070, locality 5780.

Remarks.—Two very small specimens were collected, the one from UCLA locality 5780 with its operculum in place.

Family LACUNIDAE
Genus *Boetica* Dall
Boetica hertleini Kanakoff, 1966

Boetica hertleini Kanakoff, 1966:1–4, figs. 1–3.

Hypotype.—UCLA no. 48071, locality 5758.
Age.—Early Pliocene (see remarks).

Remarks.—The type of this species was collected by Kanakoff from rocks he believed to be the upper Pliocene Pico Formation. These rocks have been mapped as Towsley Formation by Jahns and Muehlberger (1954) and are associated with a fauna very similar to that of Elsmere Canyon. This species is similar to the Recent *B. vaginata* Dall (1918:137), which lives from Monterey to La Jolla in 45 to 200 fathoms of water.

Genus *Lacuna* Turton
Lacuna sp.

Hypotype.—UCLA no. 48072, locality 5766.

Remarks.—The Tertiary and Recent species of *Lacuna* are extremely variable, and their nomenclature is consequently in a state of confusion. (See Palmer, 1958: 153–158). The present species, which is abundant at several localities in Elsmere Canyon, somewhat resembles *Lacuna carinata* Gould, but it is distinct in several features from all the known variants of that species.

On most specimens the umbilicus is closed, but some have a narrow chink at the top of the umbilical channel, which is invariably broad and shallow. This semilunar depression is bounded on the outside by a strong curved carina, which is continuous with the outer lip, and on the inside by the nearly straight columella. A very thin callus covers the inner lip from the suture halfway down the columella. The spire is low, the body whorl making up four-fifths to seven-eighths the height of the shell. Two strong angulations bound the straight sides of the body whorl, which slope at an angle of 20 degrees to the axis of the shell. The nearly horizontal shoulder makes an angle of about 110 degrees with the side, and the angle at the base of the shell is between 70 and 80 degrees. There is a broad, shallow depression on the base between the outer angulation and the umbilical carina. Earlier whorls are slightly shouldered on large specimens. Sculpture consists of very fine incremental lines which are straight and slope forward on the side of the whorl and curve back on the shoulder. The largest specimens collected are 8 mm high and 7 mm wide.

The shell figured by Dall (1921: pl. 14, fig. 2) as *Lacuna porrecta* Carpenter is also similar to these specimens, but it lacks the strongly angulated shoulder and apparently has a larger umbilicus. The very low spire and two strong angulations on the body whorl clearly distinguish this species.

Family TURRITELLIDAE
Genus *Turritella* Lamarck
Turritella cooperi Carpenter, 1864

Turritella cooperi Carpenter, 1864:612, 655; English, 1914:211; Kew, 1924:79; Grant and Gale, 1931:771–772, pl. 24, figs. 28–34.

Turritella cooperi Carpenter var. *fernandoensis* Arnold, 1907*a*:526, 538–539, pl. 51, fig. 13; Eldridge and Arnold, 1907:25, pl. 41, fig. 13.

Hypotype.—UCLA no. 48074, locality 5768.
Age.—Early Pliocene to Recent (GG:772).
Distribution.—Monterey (SG:196) to Mexico (B:54:47).
Depth.—10 to more than 60 fms (K, 1963:104; SG:196).

Family VERMETIDAE

Hypotype.—UCLA no. 48074, locality 5768.

Remarks.—A single worn specimen probably belongs in *Serpulorbis* or *Vermetus.*

Family CERITHIIDAE

Genus *Bittium* Gray

Bittium cf. *B. rugatum* Carpenter, 1864

Bittium rugatum Carpenter, 1864:539.
Bittium cf. *asperum* Gabb. English, 1914:210.
Bittium cf. *B. asperum* Gabb. Kew, 1924:78.
Bittium (*Semibittium*) *rugatum* Carpenter. Grant and Gale, 1931:762, pl. 24, fig. 8.

Hypotype.—UCLA no. 48075, locality 5758.
Age.—?Pliocene, Pleistocene to Recent (GG:762).
Distribution.—San Pedro and Catalina Island (GG:762).

Remarks.—This species is quite variable. It is thought to be most like *B. rugatum* but could not definitely be distinguished from *B. quadrifilatum* Carpenter or *B. asperum* (Gabb).

Family CALYPTRAEIDAE

Genus *Calyptraea* Lamarck

Calyptraea filosa (Gabb, 1866)

Trochita filosa Gabb, 1866 [1866–1869]:15, 81, pl. 2, figs. 25, 25*a*; Arnold, 1907*a*:527, 545, pl. 50, figs. 2, 2*a*; Eldridge and Arnold, 1907:25, pl. 40, figs. 2, 2*a*; English, 1914:210.
Calyptraea filosa Gabb. Kew, 1924:78.
Calyptraea filosa (Gabb). Grant and Gale, 1931:795.

Hypotype.—UCLA no. 48076, locality 5760.
Age.—Early Miocene to middle Pliocene (Ade:164; GG:795).

Genus *Crepidula* Lamarck

Crepidula cf. *C. aculeata* (Gmelin, 1791)

Patella aculeata Gmelin, 1791:3693.
Crepidula aculeata Gmelin. Tryon, 1886 [1879–1913]: v. 8:129, pl. 39, figs. 61–65.

Hypotype.—UCLA no. 48077, locality 5799.
Age.—Pliocene to Recent (GG:791).
Distribution.—Monterey (B:56:15) to Chile (D:162).

Remarks.—A single poorly preserved specimen was collected. *C. lingulata* Gould is lower, and its apex turns up. The apex of this specimen is higher and turns to the left, as in *C. aculeata,* but it has somewhat coarser radial ridges.

Crepidula adunca Sowerby, 1825

Crepidula adunca Sowerby, 1825: App. 7; Oldroyd, 1927: pt. 3:119, pl. 93, fig. 6.

Hypotype.—UCLA no. 48078, locality 5761.

Age.—Late Miocene (S:23) to Recent (GG:791).
Distribution.—Queen Charlotte Islands to Santo Tomas (M:35).
Depth.—Intertidal to 20 fms (SG:197; B:56:15).
Substrate.—On rocks and shells (SG:197).

Crepidula nummaria Gould, 1846

Crepidula nummaria Gould, 1846:160; Oldroyd, 1927: pt. 3:120–121, pl. 91, figs. 14, 14*a*, 14*b*.

Hypotype.—UCLA no. 48079, locality 5774.
Age.—Pliocene to Recent (GG:792).
Distribution.—Bering Strait (D:163) to Panama (B:56:16).

Remarks.—These specimens are very flat and have thinner shells than most Recent specimens.

Crepidula onyx Sowerby, 1824

Crepidula onyx Sowerby, 1824: pl. 152, fig. 2.

Hypotype.—UCLA no. 48080, locality 2056.
Age.—Late Miocene(?) to Recent (GG:791; Ade:166).
Distribution.—Southern Calif. to Chile (K, 1958:314).
Depth.—Intertidal to 50 fms (B:56:12).
Substrate.—On shells (K, 1958:314; B:56:12).

Crepidula princeps Conrad, 1855

Crepidula princeps Conrad, 1855*b*:326, pl. 6, figs. 52, 52*a*; English, 1914:210; Kew, 1924:78.

Hypotype.—UCLA no. 48081, locality 419.
Age.—Early Miocene to Pleistocene (GG:790; Ade:165).

Remarks.—This species apparently is closely related to *C. grandis* Middendorff of the Recent boreal fauna (GG:790).

Family Naticidae

Genus *Natica* Scopoli

Natica clausa Broderip and Sowerby, 1829

Natica clausa Broderip and Sowerby, 1829:360.
Natica (*Tectonatica*) *clausa* Broderip and Sowerby. Grant and Gale, 1931:797–798, text fig. 11.

Age.—Miocene to Recent (GG:797–798).
Distribution.—Arctic and Bering seas, south in deep water to San Diego (D:163).

Remarks.—This species was not collected during the present study, but a single specimen was reported from Elsmere Canyon by Grant and Gale (1931:797).

Genus *Neverita* Risso

Neverita reclusiana (Deshayes, 1839)

Natica reclusiana Deshayes, 1839:361.
Neverita recluziana Deshayes. H. and A. Adams, 1858: v. 1:208.
Neverita recluziana Petit. Arnold, 1907*a*:527; Eldridge and Arnold, 1907:25, 28, 107, 152, 153, pl. 38, fig. 6; English, 1914:211.
Natica recluziana Petit. Kew, 1924:78.
Polinices (*Neverita*) *reclusianus* (Deshayes). Grant and Gale, 1931:800–802, text figs. 13*a–c*.
Polinices (*Neverita*) *reclusianus* (Deshayes) variety *andersoni* (Clark). Grant and Gale, 1931: 802.
Polinices (*Neverita*) *reclusianus* (Deshayes) variety *pabloensis* (Clark). Grant and Gale, 1931: 802–803.

Hypotype.—UCLA no. 48082, locality 2055.
Age.—Miocene to Recent (Ade:168).
Distribution.—Mugu Lagoon (B:56:30) to Tres Marías Islands (K, 1958:324).
Depth.—Intertidal to 40 fms (Ba:708; K, 1963:102).

Genus *Polinices* Montfort
Polinices galianoi Dall, 1909

Polinices (*Euspira*) *galianoi* Dall, 1909:88–89, pl. 5, figs. 12, 13; Grant and Gale, 1931:805.

Hypotype.—UCLA no. 48083, locality 5766.
Age.—Miocene to Pleistocene (GG:805).

Remarks.—Grant and Gale (1931:805) suggested that this species might be a variety of *P. lewisii* (Gould). One specimen was collected.

Genus *Sinum* Röding
Sinum scopulosum (Conrad, 1849)

Sigaretus scopulosus Conrad, 1849. U.S. Exploring Expedition, v. 10:727, pl. 19, figs. 6, 6*a* [according to Grant and Gale, 1931:806].
Sinum scopulosum Conrad. Meek, 1864:32.
Sinum scopulosum (Conrad). Grant and Gale, 1931:806; McLean, 1969:38, fig. 19.5.

Hypotype.—UCLA no. 48084, locality 2066 (*S.* cf. *S. scopulosum*).
Age.—Oligocene to Recent (GG:806).
Distribution.—Monterey to Todos Santos Bay (as *S. californicum*—D:165; McLean 1969:38).
Depth.—Intertidal to 40 fms (K, 1963:103; B:56:32; SG:199).

Remarks.—The single specimen collected is not specifically identified, but this species has been reported from Elsmere Canyon by Grant and Gale (1931:806).

Family Cypraeidae
Genus *Cypraea* Linné
Cypraea fernandoensis Arnold, 1907

Cypraea fernandoensis Arnold, 1907*a*:526, 535, pl. 50, figs. 8, 8*a*; Eldridge and Arnold, 1907:25; English, 1914:210; Smith, 1919:153; Kew, 1924:78.
Hypotype.—UCLA no. 48085, locality 5769 (*Cypraea* sp.).
Age.—Early Pliocene (GG:752).

Remarks.—Only one unidentifiable specimen was collected during this study, but this species has been reported from Elsmere Canyon in several previous studies.

Family Cymatiidae
Genus *Cymatium* Röding
Cymatium elsmerense (English, 1914)

Gyrineum elsmerense English, 1914:215, pl. 23, fig. 6; Grant and Gale, 1931:733.
Gyrineum elsmerensis English. Kew, 1924:78.

Hypotype.—UCLA no. 48086, locality 2056.
Age.—Early Pliocene of Elsmere Canyon (GG:733).

Remarks.—The large, regular varices and flat, noded, commonly paired spiral ribs clearly place this species in the genus *Cymatium,* which is prior to Link's *Gyrineum.*

Genus *Fusitriton* Cossman
Fusitriton oregonensis (Redfield, 1846)

Triton oregonense Redfield, 1846:165–166, pl. 11, figs. 2*a*, 2*b*.

Priene (Fusitriton) oregonensis Redfield. Cossman, 1903:109, pl. 5, fig. 2.
Priene oregonensis Redfield var. *angelensis* Arnold, 1907*a*:526, 536–537, pl. 50, fig. 11; Eldridge and Arnold, 1907:25, pl. 40, fig. 11.
Ranella (Priene) oregonensis (Redfield) variety *angelensis* (Arnold). Grant and Gale, 1931:738.
Fusitriton oregonensis (Redfield, 1846). Smith, 1970:485–497, pl. 45, figs. 1–11: pl. 46, figs. 1, 2, 5, 6, 8, 9, 13, 14; pl. 47, figs. 2, 3.

Hypotype.—UCLA no. 48087, locality 5761.
Age.—Miocene to Recent (GG:737; Smith, 1970:491–492).
Distribution and depth.—Pribilof Islands to San Nicolas Island, Calif.; intertidal in Puget Sound, 26 to 330 fms in southern Oregon, and 80 to 1,300 fms in southern Calif. (Smith, 1970:491, text fig. 9).

Remarks.—This species is abundant in deep-water Pliocene rocks of the Los Angeles basin.

Family Ficidae

Genus *Ficus* Röding

Ficus (Trophosycon) ocoyana (Conrad, 1855)

Sycotypus ocoyanus Conrad, 1855*a*:19, pl. 6, fig. 72.
Ficus ocoyanus Conrad. Gabb, 1866–1869:113.
Trophosycon nodiferum Gabb. English, 1914:211; Kew, 1924:78.
Ficus nodiferus Gabb. Smith, 1919:153.
Ficus (Trophosycon) ocoyana (Conrad). Grant and Gale, 1931:743–749, pl. 29, figs. 1*a*, 1*b*, 2*a*, 2*b*; pl. 30, figs. 1–5, 7, 8*a*, 8*b*, 9*a*, 9*b*, 10*a*, 10*b*, 11.

Hypotype.—UCLA no. 48088, locality 5769.
Age.—Middle Miocene to middle Pliocene (GG:745–749; Ade:171–172).

Order NEOGASTROPODA

Family Muricidae

Genus *Forreria* Jousseaume

Forreria belcheri (Hinds, 1843)

Murex belcheri Hinds, 1843*b*:127.
Forreria belcheri Hinds. Dall, 1921:113.
Forreria magister (Nomland). Grant and Gale, 1931:727, pl. 27, figs. 14*a*, 14*b*; not *F. magister* Nomland, 1916.

Hypotype.—UCLA no. 48089, locality 5769.
Age.—Pliocene to Recent (GG:727).
Distribution.—Mugu Lagoon to Scammon's Lagoon (B:51:54).
Depth.—Intertidal to 15 fms (B:51:54)

Genus *Maxwellia* Baily

Maxwellia eldridgei (Arnold, 1907)

Murex eldridgei Arnold, 1907*a*:526, 537, pl. 50, fig. 12; Eldridge and Arnold, 1907:25, pl. 40, fig. 12; Smith, 1919:153.
Purpura (Jaton) eldridgei (Arnold). Grant and Gale, 1931:708, pl. 32, figs. 1, 2*a*, 2*b*.

Hypotype.—UCLA no. 48090, locality 5800.
Age.—Late Miocene (S:23, as *Jaton eldridgei*) to early Pliocene (GG:708).

Remarks.—This species is similar to the Recent *M. gemma* (Sowerby), which ranges from Santa Barbara to Lower Calif. (Abbott, 1954:206, pl. 24*e*). The deep pits between the varices at the suture clearly place this species in the genus *Maxwellia* (Baily, 1950:12).

Genus *Ocenebra* Gray

Ocenebra foveolata (Hinds, 1843)

Murex foveolatus Hinds, 1843*b* :127.
Murex (*Ocinebra*) *foveolatus* Hinds. Tryon, 1880 [1879–1913]:125, pl. 38, fig. 466.
Ocenebra foveolata Hinds. Burch, 1946 [1944–1946] : no. 51:49–50.

Hypotype.—UCLA no. 48091, locality 5811-1.
Age.—Early Pliocene (this report) ; Pleistocene to Recent (GG:709).
Distribution.—Monterey to Magdalena Bay (B:51:49; M:43).
Habitat.—Rocks (K, 1963:102).

Ocenebra subangulata waldorfensis Arnold, 1907

Muricidea subangulata Stearns, 1873. Conchological Memoranda no. 12:5, pl. 1, fig. 4 [according to Grant and Gale, 1931:713].
Ocinebra micheli Ford, var. *waldorfensis* Arnold, 1907*b* :435, pl. 54, fig. 10.

Hypotype.—UCLA no. 48092, locality 5811-1.
Age.—Pliocene (*O. subangulata* is Pleistocene to Recent) (GG:713).
Distribution.—*O. subangulata* ranges from Orcas Island, Wash., to Gulf of Calif. (B:51:50).

Family Thaididae

Genus *Thais* Röding

Thais elsmerensis Grant and Gale, 1931

Thais (*Nucella*) *elsmerensis* Grant and Gale, 1931:719, pl. 32, fig. 13.

Hypotype.—UCLA no. 48093, locality 5771.
Age.—Early Pliocene of Elsmere Canyon (GG:719).

Thais cf. *T. lamellosa* (Gmelin, 1791)

Buccinum lamellosum Gmelin, 1791:3498.
Thais (*Nucella*) *lamellosa* Gmelin. Dall, 1909:50.

Hypotype.—UCLA no. 48094, locality 5758.
Age.—Miocene to Recent (GG:717).
Distribution.—Bering Strait to Santa Barbara (D:111).

Family Buccinidae

Genus *Calicantharus* Clark

Calicantharus fortis angulatus (Arnold, 1907)

Pisania fortis Carpenter var. *angulata* Arnold, 1907*a*:526, 536, pl. 50, figs. 6, 7; Eldridge and Arnold, 1907:25, pl. 40, figs. 6, 7.
Pisania fortis Carpenter. Smith, 1919:153.
Cantharus fortis (Carpenter) var. *angulatus* (Arnold). Grant and Gale, 1931:674.
Calicantharus fortis angulata (Arnold). Woodring and Bramlette, 1950:48, 63, 75, pl. 14, fig. 10; pl. 15, figs. 14, 15.

Hypotype.—UCLA no. 48095, locality 5775.
Age.—*C. fortis angulatus* from late Miocene (S:23) to middle Pliocene (Ade:182); *C. fortis* Pleistocene (GG:647).

Remarks.—Some of these specimens have sculpture similar to that of typical *C. fortis*. Others have no axial sculpture at all and clearly belong in Arnold's subspecies. These are usually difficult to distinguish from *C. humerosus*.

Calicantharus humerosus (Gabb, 1869)

Neptunea humerosa Gabb, 1869 [1866–1869]:71, pl. 14, fig. 3; Arnold, 1907*a*:527; Eldridge and Arnold, 1907:25.
Chrysodomus arnoldi Rivers? Arnold, 1907*a*:526; Eldridge and Arnold, 1907:25; Kew, 1924:78.
Cantharus humerosus (Gabb). Grant and Gale, 1931:647–648, pl. 28, fig. 3.
Calicantharus humerosus (Gabb). Woodring, in Winterer and Durham, 1962:298, 312, 317, 319, 321, 322.

Hypotype.—UCLA no. 48096, locality 5760.
Age.—Pliocene to Pleistocene (GG:647).

Genus *Cantharus* Röding

Cantharus sp.

Hypotype.—UCLA no. 48097, locality 5805.

Remarks.—The single specimen is similar to several of the Recent Mexican species, including *C. elegans* (Griffith and Pidgeon), *C. ringens* (Reeve), and *C. sanguinolentus* (Duclos).

Family Neptuneidae

Genus *Kelletia* Fischer

Kelletia kelletii (Forbes, 1852)

Fusus kelletii Forbes, 1852:274, pl. 9, fig. 10.
Siphonalia kelletti Forbes. English, 1914:210.
Kelletia kellettii Forbes. Dall, 1921:89.
Siphonalia kelleti Forbes. Kew, 1924:78.
Kelletia (*Kelletia*) *kelletii* (Forbes). Grant and Gale, 1931:642–643, pl. 28, fig. 7.

Hypotype.—UCLA no. 48098, locality 5783.
Age.—Early Pliocene to Recent (GG: 643).
Distribution.—Point Conception to Asuncion Island (M:46).
Depth.—10 to 35 fms (B:50:11).
Substrate.—Offshore rocks and sand (K, 1963:101).

Genus *Neptunea* Röding

Neptunea lyrata (Gmelin, 1791)

Buccinum lyratum Gmelin, 1791:3494–3495.
Neptunea lyrata (Gmelin). Abbott, 1954:230, pl. 24*q*.

Hypotype.—UCLA no. 48099, locality 2056.
Age.—Miocene to Recent [GG:654, as *N. lirata* (Martyn)].
Distribution and depth.—Arctic Ocean to Puget Sound; off Point Pinos, Calif., in 958 fms (D:98).

Family Columbellidae

Genus *Amphissa* H. and A. Adams

Amphissa reticulata Dall, 1916

Amphissa versicolor reticulata Dall, 1916:27.

Hypotype.—UCLA no. 48100, locality 5758.
Age.—Early Pliocene (this report); Recent.
Distribution.—Port Althorp, Alaska, to San Diego (Dall, 1916:27).
Depth.—62 to 183 fms (Dall, 1916:27); 15 to 30 fms (UCLA cat. nos. 21710, 21711).

Remarks.—Dall (1916:27) suggested that this might be a distinct species. It is easily distinguished from *A. versicolor* Dall by its more acute spire and straighter axial ribs.

Genus *Mitrella* Risso

Mitrella gausapata (Gould, 1850)

Columbella gausapata Gould, 1850*a*:170.
Mitrella carinata (Hinds) variety *gausapata* (Gould). Grant and Gale, 1931:693–694, pl. 26, fig. 44.

Hypotype.—UCLA no. 48101, locality 5758.
Age.—Miocene to Recent (GG: 694).
Distribution.—700 miles north of Forrester Island, Alaska (B:51:63), to Salina Cruz, Mexico (GG:694).

Remarks.—This species does not have the distinct carina at the upper rim of the aperture which is characteristic of *M. carinata.*

Family NASSARIIDAE

Genus *Nassarius* Dumeril

Nassarius hamlini (Arnold, 1907)

Nassa hamlini Arnold, 1907*a*:537–538, pl. 50, fig. 9.
Nassarius (Catilon) hamlini (Arnold). Addicott, 1965:15, pl. 3, figs. 20, 21.

Hypotype.—UCLA no. 48103, locality 5762.
Age.—Early Pliocene (Add:15).

Nassarius iniquus (Stewart, 1940)

"Nassa" miser (Dall) var. *iniqua* Stewart, in Woodring, Stewart, and Richards, 1940:87, pl. 34, fig. 8.
Nassarius (Catilon) iniquus (Stewart). Addicott, 1965:16, pl. 3, fig. 6.

Hypotype.—UCLA no. 48104, locality 5758.
Age.—Pliocene (Add:16).

Remarks.—Addicott (1965:16) described a weakly noded and more elongate variant from Elsmere Canyon. The specimens from Elsmere Canyon are variable, however, ranging from a few slender elongate individuals to the more abundant obese form typical of the species. The slender form is more abundant at the top of the lower unit and in the upper unit of the Towsley Formation in Elsmere Canyon.

Nassarius stocki Kanakoff, 1956

Nassarius stocki Kanakoff, 1956:110–112, pl. 30, figs. A–C.

Hypotype.—UCLA no. 48105, locality 5758.
Age.—Late Miocene to early Pliocene (Add:14).

Nassarius whitneyi (Trask, 1922)

Nassa whitneyi Trask, 1922:154, pl. 7, figs. 3, 6.
Nassarius whitneyi Trask. Clark, 1929: pl. 34, fig. 9.

Hypotype.—UCLA no. 48106, locality 5786.
Age.—Early Pliocene (this report); middle to late Miocene (Add:10).

Remarks.—At Elsmere Canyon this species occurs only in the basal beds.

Family MITRIDAE

Genus *Mitra* Röding

Mitra idae Melvill, 1893

Mitra idae Melvill, 1893:140–141, pl. 1, fig. 6; Grant and Gale, 1931:635.

Mitra idae Dall. English, 1914:211; Kew, 1924:78.

Hypotype.—UCLA no. 48107, locality 5758.
Age.—Early Pliocene to Recent (GG:635).
Distribution.—Crescent City to Cedros Island (M:50).
Depth.—Intertidal to 10 fms (B:49:30).
Substrate.—Rocks (K, 1963:102).

Family Fusinidae
Genus *Fusinus* Rafinesque, 1815
Fusinus barbarensis (Trask, 1855)

Fusus barbarensis Trask, 1855:41; English, 1914:211.
Fusinus barbarensis Trask. Kew, 1924:78; Dall, 1921:88.
Fusinus barbarensis (Trask). Grant and Gale, 1931:639, pl. 27, fig. 1.

Age.—Pliocene to Recent (GG:639).
Distribution.—Heceta Bank, Oregon, to San Diego (GG:639).

Remarks.—This species was not collected during the present study but was reported from Elsmere Canyon by English (1914:211).

Family Olividae
Genus *Olivella* Swainson
Olivella aff. *O. gracilis* (Broderip and Sowerby, 1829)

Oliva gracilis Broderip and Sowerby, 1829:379.
Olivella gracilis Broderip and Sowerby. Tryon, 1883 [1879–1913]: v. 5:70, pl. 16, fig. 26.

Hypotype.—UCLA no. 48108, locality 5785.
Age.—Pliocene to Recent (GG:627).
Distribution.—Guaymas to Peru (K, 1958:424).

Remarks.—Poorly preserved specimens from two localities are similar to *O. Gracilis,* which occurs in Pliocene rocks of Imperial County but is not known, Recent or fossil, from the outer coast.

Olivella pedroana (Conrad, 1855)

Strephona pedroana Conrad, 1855*a*:17.
Olivella pedroana Conrad. Arnold, 1903:221.
Olivella intorta Carpenter. Arnold, 1907*a*:527; Eldridge and Arnold, 1907:25.
Olivella pedroana (Conrad). Grant and Gale, 1931:626–627, pl. 24, fig. 10.

Hypotype.—UCLA no. 48109, locality 5786.
Age.—Middle Miocene to Recent (GG:626–627; Ade:188).
Distribution.—Puget Sound to Cape San Lucas (D:85).
Depth.—Intertidal to 40 fms (B:49:21; P:341).

Family Cancellariidae
Genus *Admete* Moller
Admete rhyssa Dall, 1919

Admete rhyssa Dall, 1919:306.

Hypotype.—UCLA no. 48110, locality 5758.
Age.—Early Pliocene (this report); Pleistocene to Recent (GG:623).
Distribution.—Santa Rosa Island to S. Coronado Island (D:84).

Genus *Cancellaria* Lamarck

Cancellaria arnoldi Dall, 1909

Cancellaria arnoldi Dall, 1909:29–30, pl. 14, fig. 7.

Hypotype.—UCLA no. 48111, locality 2066.
Age.—Pliocene (Dall, 1909:30).

Cancellaria elsmerensis English, 1914

Cancellaria elsmerensis English, 1914:216, pl. 23, fig. 8; Kew, 1924:78, Grant and Gale, 1931: 616.

Hypotype.—UCLA no. 48112, locality 5779.
Age.—Early Pliocene of Elsmere Canyon (GG:616).

Cancellaria fernandoensis Arnold, 1907

Cancellaria fernandoensis Arnold, 1907*a*:526, 535, pl. 50, fig. 4; English, 1914:210; Kew, 1924:78; Eldridge and Arnold, 1907:25, pl. 40, fig. 4.
Cancellaria tritonidea Gabb variety *fernandoensis* Grant and Gale, 1931:618–619, pl. 27, fig. 1.

Hypotype.—UCLA no. 48113, locality 5758.
Age.—Late Miocene (questionable—S:23); early to late Pliocene (GG:618).

Remarks.—Grant and Gale (1931:618–619, pl. 27, fig. 1) considered *C. fernandoensis* a subspecies of *C. tritonidea* Gabb. See "remarks" under that species.

Cancellaria hamlini Carson, 1926

Cancellaria hamlini Carson, 1926:51, pl. 1, figs. 4, 6.
Cancellaria tritonidea Gabb variety *hamlini* Carson. Grant and Gale, 1931:617–618.

Hypotype.—UCLA no. 48114, locality 5779.
Age.—Early Pliocene of Elsmere Canyon (Carson, 1926:51).

Remarks.—Grant and Gale considered *C. hamlini* a subspecies of *C. tritonidea* Gabb. It differs from *C. rapa* in being strongly, but roundly, shouldered and having more delicate, but more elevated, sculpture. The columellar callus does not extend above the body whorl onto the outside of earlier whorls, as it commonly does in *C. tritonidea*, *C. fernandoensis*, and *C. rapa*.

Cancellaria planospira Grant and Gale, 1931

Cancellaria obesa Sowerby variety *Planospira* Grant and Gale, 1931:613, pl. 27, fig. 4.

Hypotype.—UCLA no. 48115, locality 5782.
Age.—Early Pliocene of Elsmere Canyon (GG:613).

Remarks.—This species has a shape different from any of the forms of the variable *C. obesa.*

Cancellaria rapa Nomland, 1917

Cancellaria rapa Nomland, 1917:240, pl. 11, figs. 1, 1*a*; Smith, 1919:153.
Cancellaria tritonidea Gabb variety *rapa* Nomland. Grant and Gale, 1931:617.

Hypotype.—UCLA no. 48116, locality 5760.
Age.—Late Miocene (S:23) to late Pliocene (GG:617; Ade:191).

Remarks.—Grant and Gale (1931:617) included *C. rapa* as a subspecies of *C. tritonidea* Gabb. See "remarks" under that species.

Cancellaria tritonidea Gabb, 1869

Cancellaria (*Euclia*) *tritonidea* Gabb, 1869 [1866–1869]:11, 79, pl. 2, fig. 18.
Cancellaria tritonidea Gabb. English, 1914:210; Smith, 1919:153; Kew, 1924:78; Grant and Gale, 1931:616–617, pl. 27, fig. 8.

Hypotype.—UCLA no. 48117, locality 5771.
Age.—Late Miocene (S:23, 34); early Pliocene to Pleistocene (GG:616–617).

Remarks.—Four of Grant and Gale's six subspecies have been collected in this study. *C. hamlini* is considered definitely a distinct species, though possibly related. The other three have been listed here as different species—*C. tritonidea, C. fernandoensis,* and *C. rapa*—although not without some reservation. Although *C. tritonidea* seems to be intermediate between the other two, I feel that any specimen could be decisively placed in one of the three groups. *C. fernandoensis* has very strong, regular axial and spiral sculpture on all the whorls and does not have elevated nodes on the nearly squared shoulders. *C. rapa* has very irregular and often weak sculpture, especially on the early whorls. It has no shoulder; the whorls are regularly rounded. Very small nodes may be present on the latest whorls. *C. tritonidea* has fairly regular sculpture, though not so strong as in *C. fernandoensis,* and the early whorls are commonly similar to those of *C. rapa.* Large, elevated nodes form on the shoulder, commonly even on the early whorls, and the shoulder above the whorls commonly is steeply sloping. All three forms may have columellar callus extending above the body whorl to cover the outside of the previous whorl, though not so commonly in *C. fernandoensis* as in the other two. Whatever interpretation is accepted, it is clear that these forms are very closely related.

Family Marginellidae

Genus *Marginella* Lamarck

Marginella sp.

Hypotype.—UCLA no. 48102, locality 5758.

Remarks.—The single poorly preserved specimen appears to be unlike other west coast species of *Marginella.*

Family Conidae

Genus *Conus* Linné

Conus californicus Reeve, 1844

Conus californicus Reeve, in Hinds, 1844: pl. 1, figs. 3–5.
Conus californicus Hinds. English, 1914:210; Kew, 1924:78; Grant and Gale, 1931:472, pl. 24, fig. 21.

Hypotype.—UCLA no. 48118, locality 5758.
Age.—Late Miocene (S:23) to Recent (GG:472).
Distribution.—Farallon Islands to Ballenas Lagoon (D:68).
Depth.—Intertidal to 25 fms (B:48:23; RC:145; SG: 182).
Substrate.—Sand and gravel, rocks (B:48:23; RC:145; SG:182).
Habitat.—Estuaries and offshore (B:48:23).

Family Terebridae

Genus *Terebra* Bruguiere

Terebra martini English, 1914

Terebra martini English, 1914:216, pl. 23, fig. 3.

Terebra (Strioterebrum) elata **Hinds variety** ***martini*** **English. Grant and Gale, 1931:470.**

Hypotype.—UCLA no. 48119, locality 5758.

Age.—Early Pliocene of Elsmere Canyon (GG:470).

Remarks.—Grant and Gale (1931:470) included this taxon as a subspecies of *T. elata* Hinds, but it lacks the spiral sculpture of that species.

Family TURRIDAE

Genus *Clathrodrillia* Dall

Clathrodrillia coalingensis (Arnold, 1909)

Pleurotoma coalingensis **Arnold, 1909:90, pl. 22, fig. 2.**

Turris elsmerensis **English, 1914:211, 216–217, pl. 23, figs. 4*a*, 4*b*; Kew, 1924:79.**

Clavus (Clathrodrillia) coalingensis **(Arnold). Grant and Gale, 1931:580, pl. 26, figs. 14*a*, 14*b*, 15.**

Hypotype.—UCLA no. 48120, locality 2055.

Age.—Late Miocene (S) to middle Pliocene (GG:580; Ade:195).

Remarks.—This species has the well-developed evenly expressed axial and spiral sculpture of the genus *Clathrodrillia.* The spiral sculpture of *Clavus* is less well developed or lacking (K, 1958:448, 453).

Genus *Clavus* Montfort

Clavus aff. *C. pallidus* (Sowerby, 1834)

Pleurotoma pallida **Sowerby, 1834*a*:137.**

Clavis (Cymatosyrinx) pallidus **(Sowerby). Grant and Gale, 1931:576–577, pl. 26, figs. 16*a*, 16*b*, 17.**

Hypotype.—UCLA no. 48121, locality 5758.

Age.—Early Pliocene (this report); middle Pliocene to Recent (GG:577).

Distribution.—Santa Barbara to Panama (GG:577); Tenacatita Bay, Mexico, to Colombia (K, 1958:452).

Genus *Crassispira* Swainson

Crassispira aff. *C. montereyensis* (Stearns, 1871)

Pleurotoma (Drillia) montereyensis **Stearns, 1871:2.**

Crassispira montereyensis **Stearns. Dall, 1921:70.**

Clavus (Crassispira) montereyensis **(Stearns). Grant and Gale, 1931:581–582, pl. 26, fig. 9.**

Hypotype.—UCLA no. 48122, locality 5758.

Age.—Pleistocene to Recent (GG:582).

Distribution.—Monterey to Mazatlan (D:70).

Remarks.—These specimens differ from Recent specimens of this species in having much weaker sculpture on the subsutural band.

Genus *Mangelia* Risso

Mangelia interlirata Stearns, 1872

Mangelia interlirata **Stearns, 1872:226, pl. 1, fig. 10.**

Hypotype.—UCLA no. 48123, locality 5758.

Age.—Early Pliocene (this report); late Pliocene to Recent (GG:592).

Distribution.—Monterey (GG:592) to San Martin Island (Baker, 1902:41).

Genus *Spirotropis* Sars

Spirotropis perversa (Gabb, 1865)

Pleurotoma (Surcula) perversa **Gabb, 1865:183.**

Turris fernandoensis English, 1914:211, 217, pl. 23, fig. 7; Kew, 1924:79.
Spirotropis (*Antiplanes*) *perversa* (Gabb) variety *fernandoensis* (English). Grant and Gale, 1931:557, pl. 26, fig. 25.

Age.—Late Miocene to Recent (GG:557).
Distribution.—Alaska to San Diego and Cortez Bank (GG:554).

Remarks.—This species was not collected during the present study but has been reported from Elsmere Canyon by English (1914:217) and Grant and Gale (1931: 556).

Genus *Megasurcula* Casey

Megasurcula carpenteriana (Gabb, 1865)

Pleurotoma (*Surcula*) *carpenteriana* Gabb, 1865:183–184.
Bathytoma cf. *carpenteriana* Gabb. Arnold, 1907*a*:526.
Bathytoma carpenteriana Gabb, var. *fernandoana* Arnold, 1907*b*:432–433, pl. 56, fig. 7; English, 1914:210.
Dolichotoma cf. *carpenteriana* Gabb. Eldridge and Arnold, 1907:25.
Turris (*Bathytoma*) *carpenteriana* Gabb var. *fernandoana* Arnold. Kew, 1924:79.
Surculites (*Megasurcula*) *carpenteriana* (Gabb) variety *cooperi* (Arnold). Grant and Gale, 1931:499–500, pl. 25, fig. 3.
Megasurcula carpenteriana (Gabb). Burch, 1946 [1944–1946]: no. 62:5.

Hypotype.—UCLA no. 48124, locality 5782.
Age.—Pliocene to Recent (GG:498–500).
Distribution.—Bodega Bay (D:68) to Cedros Island (B:62:5).
Depth.—12 to more than 200 fms (SG:182; P:344; B:62:5).
Substrate.—Mud and fine sand (SG:182; P:344; B:62:5).

Genus *Pseudomelatoma* Dall

Pseudomelatoma aff. *P. moesta* (Carpenter, 1864)

Drillia moesta Carpenter, 1864:537, 657; Tryon, 1884 [1879–1913]: v. 6:183, pl. 12, fig. 38.
Pseudomelatoma moesta Carpenter. Dall, 1921:70.

Hypotype.—UCLA no. 48125, locality 5760.
Age.—Late Pleistocene to Recent (GG:561).
Distribution.—Monterey to Cedros Island (D:70).

Remarks.—Coarse axial costae and fine spiral grooves are better developed than on Recent specimens.

Genus *Ophiodermella* Bartsch

Ophiodermella cancellata (Carpenter, 1864)

Drillia cancellata Carpenter, 1864:603, 658.
Moniliopsis rhines Dall, 1919:28, pl. 8, fig. 5, new name for *Drillia cancellata* Carpenter.
Ophiodermella rhines (Dall). Burch, 1946 [1944–1946]: no. 62:10.

Hypotype.—UCLA no. 48126, locality 5800.
Age.—Early Pliocene (this report); Recent (Palmer, 1958:226).
Distribution.—Puget Sound to San Diego (D:70; Palmer, 1958:226).

Remarks.—Carpenter (1864:657, 658) named two species of *Drillia: D. incisa,* with "spiral sculpture grooved, not raised," and "nodosely cancellated" *D. cancellata.* Grant and Gale (1931:565–566) incorrectly placed the two in synonymy as *D. incisa,* in spite of the different sculpture. Dall (1919:28, pl. 8, fig. 5) proposed *Moniliopsis rhines* as a new name for *Drillia cancellata,* incorrectly believing that that name was preoccupied (Palmer, 1958: 226). The two specimens collected during this study are the same as the species figured by Dall, which has cancellate sculpture. If Dall's *Moniliopsis rhines* is the same as Carpenter's *Drillia cancellata,* of which the type has not been found, the present specimens are correctly referred to *Ophiodermella cancellata.*

Genus *Taranis* Jeffreys

Taranis incultus (Moody, 1916)

Borsonia inculta Moody, 1916:54–55, pl. 1, figs. 2*a*, 2*b*.
Taranis incultus (Moody). Grant and Gale, 1931:572.

Hypotype.—UCLA no. 48127, locality 5758.
Age.—Early Pliocene (this report); late Pliocene (GG:572).

Remarks.—This species is similar to *T. strongi* (Arnold) but has much weaker spiral sculpture.

Subclass OPISTHOBRANCHIATA

Order CEPHALASPIDEA

Family Bullidae

Genus *Bulla* Linné

Bulla cf. *B. punctulata* A. Adams, 1850

Bulla punctulata A. Adams, in Sowerby, 1850 [1847–1887]: v. 2:577–578 [as *B. punctata* by mistake, not *B. punctata* Schroeter, corrected on p. 604], pl. 12, fig. 77; English, 1914:210.

Hypotype.—UCLA no. 48128, locality 5786.
Age.—Pliocene to Recent (GG:456).
Distribution.—Magdalena Bay and Gulf of Calif. to Peru (K, 1958: 496).
Depth.—Offshore beyond low tide (K, 1958:496).

Family Pyramidellidae

Genus *Turbonilla* Risso

Turbonilla sp.

Hypotype.—UCLA no. 48129, locality 5758.

Family Scaphandridae

Genus *Acteocina* Gray

Acetocina aff. *A. culcitella* (Gould, 1853)

Bulla (*Akera*) *culcitella* Gould, 1853:377–378, pl. 14, fig. 8.
Acteocina culcitella Gould. Dall, 1921:61.

Hypotype.—UCLA no. 48130, locality 5786.
Age.—Pliocene to Recent (GG:447).
Distribution.—Kodiak Island (D:61) to Scammon's Lagoon (J).

Remarks.—These specimens are similar to *A. culcitella* but are much smaller; most of the specimens are less than 6 mm long.

Family Acteonidae

Genus *Acteon* Montfort

Acteon punctocaelatus (Carpenter, 1864)

Tornatella punctocaelata Carpenter, 1864:646.
Actaeon (*Rictaxis*) *puncto-coelata Carpenter.* Dall, 1871:136, 160, pl. 15, fig. 12.
Acteon punctocaelatus Carpenter. Oldroyd, 1914:81.

Hypotype.—UCLA no. 48131, locality 5786.
Age.—Early Pliocene (this report); Pleistocene to Recent (GG:443).
Distribution.—British Columbia to Magdalena Bay (B:47:9).
Depth.—3 to 50 fms (B:47:10; Ba; P:345).
Substrate.—Sand or mud (K, 1963:99; L:366; RC:261).

Class SCAPHOPODA
Family Dentaliidae
Genus *Dentalium* Linné
Dentalium neohexagonum Sharp and Pilsbry, 1897

Dentalium neohexagonum Sharp and Pilsbry, in Tryon, 1897 [1879–1913]: v. 17:19–20, pl. 11, figs. 74–86.

Hypotype.—UCLA no. 48132, locality 5758.
Age.—Pliocene to Recent (GG:436).
Distribution.—Monterey to Central America (D:57).

Remarks.—The few fragmentary specimens have from seven to ten weak longitudinal ridges.

Dentalium cf. *D. rectius* Carpenter, 1864

Dentalium rectius Carpenter, 1864:603, 648; Oldroyd, 1927: pt. 1:11, pl. 1, fig. 3.

Hypotype.—UCLA no. 48133, locality 5800.
Age.—Miocene(?), Pliocene to Recent (GG:437).
Distribution.—Stephens Passage, Alaska, to Panama Bay (D:57).

Remarks.—These specimens are straight, round, and very thin-shelled.

Phylum ARTHROPODA
Class CRUSTACEA
Subclass CIRRIPEDIA
?*Tetraclita* sp.

Remarks.—These barnacles have four compartments as in *Tetraclita:* a carina, two laterals, and a fused rostrum-rostrolaterals. Several specimens also retain the original reddish color. The cavities in the wall are narrow and radially elongate (with respect to the animal), rather than round in section like those of *Tetraclita.*

Phylum ECHINODERMATA
Class ECHINOIDEA
Order CLYPEASTEROIDA
Family Echinarachniidea
Genus *Astrodapsis* Conrad
Astrodapsis fernandoensis Pack, 1909

Astrodapsis fernandoensis Pack, 1909:279, pl. 24, figs. 3, 4; English, 1914:209, pl. 23, fig. 5; Smith, 1919:152; Kew, 1924:77; Grant and Hertlein, 1938:72–73, pl. 25, figs. 4, 5.

Hypotype.—UCLA no. 48134, locality 5799.
Age.—Late Miocene (S:23) to early Pliocene (Hall, 1962:77).

Family Dendrasteridae
Genus *Dendraster* L. Agassiz
Dendraster elsmerensis Durham, 1949

Dendraster excentricus Eschscholtz. Pack, 1909:281–282; English 1914:209; not *D. excentricus* Eschscholtz, 1831.

Echinarachnius excentricus Eschscholtz. Kew, 1924:77; not *D. excentricus* Eschscholtz, 1831.
Dendraster elsmerensis Durham, 1949:50–62, pl. 1, figs. 2–4, 6.

Age.—Early Pliocene of Elsmere Canyon (Durham, 1949:50–62).

Remarks.—This species was not collected during the present study.

Dendraster gibbsii Remond, 1863

Scutella gibbsii Remond, 1863:13–14.
Dendraster gibbsii Remond. Clark and Twitchell, 1915: 193–195, pl. 89, figs. 1–4; pl. 108, fig. B.
Dendraster ashleyi Arnold. Smith, 1919:152; not *D. ashleyi* Arnold, 1907.

Hypotype.—UCLA no. 48135, locality 5797.
Age.—Miocene to Pliocene (Grant and Hertlein, 1938:88).

Phylum CHORDATA
Subphylum VERTEBRATA
Class CHONDRICHTHYES
Family LAMNIDAE
Genus *Carcharodon* Muller and Henle
Carcharodon megalodon Agassiz, 1843

Hypotype.—UCLA no. 48137, locality 5771.
Age.—Oligocene to Pleistocene (S. P. Applegate, oral communication, 1967).

Remarks.—This and the following two species were identified by Dr. Shelton Applegate of the Los Angeles County Museum.

Carcharodon sulcidens Agassiz, 1843

Hypotype.—UCLA no. 48136, locality 5758.
Age.—Oligocene to Recent (S. P. Applegate, oral communication, 1967).

Genus *Isurus* Rafinesque
Isurus hastalis (Agassiz, 1843)

Hypotype.—UCLA no. 48138, locality 5771.
Age.—Oligocene to Recent (S. P. Applegate, oral communication, 1967).

Class MAMMALIA

Remarks.—A single whale vertebra and numerous unidentified bone fragments were collected.

Trace Fossils

Remarks.—Burrows in sand were collected at three localities, and mollusc shells are bored by unidentified organisms at six localities.

Plants

Remarks.—Unidentified plant fossils include parts of a leaf and a cone and fragments of wood at a number of localities.

LOCALITY DESCRIPTIONS

Numbers refer to localities described in the invertebrate paleontology locality catalogue, University of California, Los Angeles. The localities are all in the lower Pliocene Towsley Formation in T. 3 N., R. 15 W., U.S. Geological Survey 7.5-minute San Fernando quadrangle (1953) (fig. 4).

L419

2,750 feet west and 760 feet north of southeast corner of section 7. Coarse-grained sandstone and conglomerate overlying Eocene rocks in cliff on south side of canyon.

L2055

440 feet east and 2,250 feet north of southwest corner of section 8. Calcareous silty sandstone of fossil bed A, 20 feet above stream just north of hanging valley.

L2056

4,340 feet west and 2,810 feet north of southeast corner of section 7. Calcareous coarse-grained sandstone of old well site 15 feet up south slope of canyon.

L2066

670 feet east and 1,000 feet north of southwest corner of section 17. Northernmost of several prominent knobs of calcareous sandstone.

5758

2,650 feet west and 240 feet north of southeast corner of section 18. Thin bed of coarse-grained sandstone in northeast-facing cliff of siltstone.

5759

2,600 feet west and 190 feet north of southeast corner of section 18. Thin bed of sandstone exposed 2 feet below upper lip of deep narrow gully on east side of road.

5760

2,630 feet west and 375 feet south of northeast corner of section 19. Sandy bed in siltstone, poorly exposed near top of steep west slope of Grapevine Canyon.

5761

2,630 feet west and 535 feet south of northeast corner of section 19. Same bed as at 5760.

5762

1,480 feet west and 1,470 feet south of northeast corner of section 19. Thin sandstone bed in siltstone, 200 feet southeast of Edison Co. tower.

5763

2,020 feet west and 100 feet south of northeast corner of section 19. Calcareous sandy siltstone in road cut.

5764

1,800 feet west and 70 feet north of southeast corner of section 18. Calcareous sandstone float in cut made for gas pipeline.

5765

1,880 feet west and 330 feet north of southeast corner of section 18. Platy calcareous sandstone float just west of south end of narrow cut made for gas pipeline.

5766

1,880 feet west and 430 feet north of southeast corner of section 18. Diagonally up slope from 5765.

5767

1,800 feet west and 480 feet north of southeast corner of section 18. Similar to 5765.

5768

1,480 feet west and 550 feet north of southeast corner of section 18. Basal conglomerate on north bank of stream.

5769

1,130 feet west and 780 feet north of southeast corner of section 18. Basal conglomerate on south bank of stream.

5770

1,210 feet west and 120 feet south of northeast corner of section 19. Concretions in coarse-grained basal sandstone on north rim of canyon.

5771

770 feet west and 30 feet north of southeast corner of section 18. Same rock as at 5770, on north wall of canyon.

5772

420 feet west and 130 feet south of northeast corner of section 19. Calcareous sandstone on steep south bank of small brushy gully.

5773

150 feet east and 50 feet south of northwest corner of section 20. Southernmost of several prominent knobs of calcareous sandstone. (See localities L2066, 5774–5776.)

5774

350 feet east and 550 feet north of southwest corner of section 17. Same rock as 5773.

5775

550 feet east and 750 feet north of southwest corner of section 17. Same rock as 5773.

5776

470 feet east and 830 feet north of southwest corner of section 17. Same rock as 5773.

5777

350 feet east and 770 feet north of southwest corner of section 17. Fossils disseminated in bituminous sandy siltstone on slope northwest of stream.

5778

300 feet west and 1,650 feet north of southeast corner of section 18. Concretionary bed in siltstone, 15 feet above basement contact on north side of canyon.

5779

350 feet west and 1,750 feet north of southeast corner of section 18. Conglomerate lens 25 feet above stream on north side of canyon.

5780

2,000 feet west and 2,300 feet north of southeast corner of section 18. Very small lens (1 foot by 3 feet) of conglomeratic siltstone in sandstone-conglomerate tongue. About 60 feet below contact with bituminous rocks of Pico Formation on very steep slope.

5781

800 feet west and 3,000 feet north of southeast corner of section 18. Fossils weathered out of soft siltstone on rim of north slope of canyon.

5782

580 feet west and 3,040 feet north of southeast corner of section 18. Fossils weathered out of thin bed in siltstone near crest of ridge.

5783

150 feet east and 2,450 feet north of southwest corner of section 17. Basal calcareous sandstone on brushy north rim near head of canyon.

5784

100 feet east and 3,380 feet north of southwest corner of section 17. Basal coarse-grained bituminous sandstone.

5785

640 feet west and 180 feet south of northeast corner of section 18. Basal conglomerate 50 feet above stream on southwest wall of canyon.

5786

580 feet west and 90 feet north of southeast corner of section 7. Basal conglomerate on northeast side of canyon bottom.

5787

370 feet west and 350 feet north of southeast corner of section 7. Basal conglomerate on northwest rim of canyon.

5788

420 feet west and 330 feet north of southeast corner of section 7. Basal conglomerate 35 feet above stream on northwest wall of canyon.

5789

780 feet west and 440 feet north of southeast corner of section 7. Basal conglomerate about 25 feet above stream on shelf on north wall of canyon.

5790

930 feet west and 250 feet north of southeast corner of section 7. Two concretionary beds in low cliff of siltstone 50 feet above stream on steep south slope of canyon.

5791

1,350 feet west and 230 feet north of southeast corner of section 7. 50 feet northwest from prospect pit. Concretions and bituminous sandy siltstone 20 feet above stream on north wall of canyon.

5792

800 feet west and 680 feet north of southeast corner of section 7. Calcareous sandy siltstone near top of ridge.

5793

500 feet west and 580 feet north of southeast corner of section 7. Calcareous sandstone on top of ridge.

5794

560 feet west and 980 feet north of southeast corner of section 7. Calcareous sandy siltstone of bed A on northwest side of shallow canyon. This bed is continuously fossiliferous from here to 5795.

5795

140 feet west and 950 feet north of southeast corner of section 7. Same bed as at 5794.

5796

160 feet east and 1,720 feet north of southwest corner of section 8. Basal coarse-grained bituminous sandstone at fork of canyon.

5797

180 feet east and 1,770 feet north of southwest corner of section 8. Calcareous sandstone in canyon bottom.

5798

430 feet east and 1,730 feet north of southwest corner of section 8. Basal calcareous sandstone in canyon bottom.

5799

660 feet east and 1,830 feet north of southwest corner of section 8. Calcareous coarse-grained sandstone just south of ridge crest.

5800

640 feet east and 1,910 feet north of southwest corner of section 8. Calcareous sandy siltstone of bed A exposed on crest of ridge.

5800–1

Bed C crops out 80 feet west of 5800.

5800–2

Bed D crops out 40 feet west of 5800–1.

5801

730 feet east and 2,350 feet north of southwest corner of section 8. Calcareous silty sandstone of Bed A, 20 feet above stream on northwest wall of canyon.

5801–1

Bed B crops out 10 feet above 5801.

5802

440 feet east and 2,250 feet north of southwest corner of section 8. Calcareous sandy siltstone of Bed A, 20 feet above stream on northwest wall of canyon.

5803

360 feet east and 2,250 feet north of southwest corner of section 7. Four fossil beds exposed on steep northwest wall of canyon; lowest is bed C, 35 feet above stream.

5803–1

Bed D, 8 feet above C.

5803–2

Bed E, 12 feet above D.

5803–3

Bed G, 28 feet above E.

5804

440 feet east and 2,480 feet north of southwest corner of section 8. Two fossil beds exposed on east side of canyon; lowest is bed D.

5804–1

Bed E, 12 feet above D.

5805

120 feet west and 2,430 feet north of southeast corner of section 7. Silty sandstone in sandstone-conglomerate tongue. Small ledge on steep grassy slope.

5806

310 feet east and 2,530 feet north of southwest corner of section 8. Sandstone above thin bed of conglomerate 15 feet above stream on west side of canyon.

5807

120 feet east and 2,340 feet north of southwest corner of section 8. Calcareous sandy siltstone below lip of dry falls on north side of canyon.

5808

120 feet east and 2,130 feet north of southwest corner of section 8. Two fossil beds are accessible on sloping shelf about 25 feet above stream on northwest side of canyon; lower is bed B.

5808–1

Bed C is 10 feet above B.

5809

70 feet east and 2,070 feet north of southwest corner of section 8. Seven fossil beds are exposed on northwest wall of canyon; lowest is bed A, 15 feet above stream.

5809–1

Bed B is 20 feet above A.

5809–2

Bed C is 8 feet above B.

5809–3

Bed D is 8 feet above C.

5809–4

Bed E is 12 feet above D.

5809–5

Bed F is 10 feet above E.

5809–6

Bed G is 14 feet above F.

5810

30 feet west and 1,950 feet north of southeast corner of section 7. Two fossil beds are exposed on northwest wall of canyon. Lower is bed A, 15 feet above stream.

5810–1

Bed G is 80 feet above A.

5811

210 feet west and 1,800 feet north of southeast corner of section 7. Bed A is 15 feet above stream.

5811–1

Bed G is 80 feet above A.

5812

570 feet west and 1,370 feet north of southeast corner of section 7. Calcareous sandy siltstone of bed A, 30 feet above stream on northwest wall of canyon.

5813

740 feet west and 1,510 feet north of southeast corner of section 7. Coarse-grained sandstone bed in siltstone, 75 feet above stream on northwest rim of canyon.

5814

810 feet west and 1,450 feet north of southeast corner of section 7. Possibly same bed as at 5813. Northwest rim of canyon.

5815

800 feet west and 1,200 feet north of southeast corner of section 7. Calcareous silty sandstone of bed A, 20 feet above stream on north wall of canyon.

5816

1,380 feet west and 1,180 feet north of southeast corner of section 7. Few fossils in concretions in siltstone at base of cliff of sandstone and conglomerate.

5817

2,590 feet west and 40 feet north of southeast corner of section 7. Basal coarse-grained sandstone and conglomerate near contact with Eocene sandstone.

5818

1,650 feet west and 780 feet south of northeast corner of section 18. Calcareous sandstone on crest of ridge northwest of saddle.

LITERATURE CITED

ABBOTT, R. T.

1954. American seashells. Toronto, New York, London: D. Van Nostrand. 541 pp.

ADAMS, H., and A. ADAMS

1858. The genera of Recent Mollusca. London. Vol. 1, 484 pp.; vol. 2, 661 pp.

ADDICOTT, W. O.

1965. Some western American Cenozoic gastropods of the genus *Nassarius*. U.S. Geol. Survey, Prof. Paper 503-B, 24 pp., pls. 1–3.

1968. Neogene molluscan zoogeography and climatic change in the northeastern Pacific Ocean (abstract). Geol. Soc. Amer. Program, Annual Meeting, pp. 2–3.

1969*a*. Tertiary climatic change in the marginal northeastern Pacific Ocean. Science, 165:583–586.

1969*b*. Late Pliocene mollusks from San Francisco peninsula, California, and their paleogeographic significance. Calif. Acad. Sci., Proc., ser. 4, 37 (3) :57–93.

1970*a*. Tertiary paleoclimatic trends in the San Joaquin basin, California. U.S. Geol. Survey, Prof. Paper 644-D, 19 pp.

1970*b*. Latitudinal gradients in Tertiary molluscan faunas of the Pacific coast. Palaeogeog., Palaeoclimatol., Palaeoecol., 8:287–312.

ADDICOTT, W. O., and W. K. EMERSON

1959. Pleistocene invertebrates from Punta Cabras, Baja California, Mexico. Amer. Mus. Novitates, no. 1925:1–33.

ADDICOTT, W. O., and J. G. VEDDER

1963. Paleotemperature inferences from late Miocene mollusks in the San Luis Obispo–Bakersfield area, California. U.S. Geol. Survey, Prof. Paper 475-C:63–68.

ADEGOKE, O. S.

1969. Stratigraphy and paleontology of the marine Neogene formations of the Coalinga region, California. Univ. Calif. Publ. Geol. Sci., 80, 241 pp., 13 pls.

ALLAN HANCOCK FOUNDATION

1965. An oceanographic and biological survey of the southern California mainland shelf. Calif. State Water Quality Control Board, Publ. no. 27, 232 pp.

APPELLÖF, A.

1912. Invertebrate bottom fauna of the Norwegian Sea and North Atlantic. *In* The depths of the ocean, ed. J. Murray and J. Hjort. Pp. 457–560. London: Macmillan.

ARNOLD, RALPH

1903. Paleontology and stratigraphy of the marine Pliocene and Pleistocene of San Pedro, California. Calif. Acad. Sci., Mem., 3:1–420, 37 pls.

1906. The Tertiary and Quaternary Pectens of California. U.S. Geol. Survey, Prof. Paper 47, 264 pp., 53 pls.

1907*a*. New and characteristic species of fossil mollusks from the oil-bearing Tertiary formations of southern California. U.S. Natl. Mus., Proc., 32:525–546, pls. 38–51.

1907*b*. New and characteristic species of fossil mollusks from the oil-bearing Tertiary formations of Santa Barbara County, California. Smithsonian Misc. Coll., 50 (pt. 4) :419–447.

1909. Paleontology of the Coalinga district, Fresno and Kings counties, California. U.S. Geol. Survey, Bull. 396, 173 pp.

ASHLEY, G. H.

1895. The Neocene stratigraphy of the Santa Cruz Mountains of California. Calif. Acad. Sci., Proc., ser. 2, 5:273–367.

AXELROD, D. I.

1941. The concept of ecospecies in Tertiary paleobotany. Natl. Acad. Sci., Proc., 27:545–551.

1956. Mio-Pliocene floras from west-central Nevada. Univ. Calif. Publ. Geol. Sci., 33:1–327.

1965. A method for determining the altitudes of Tertiary floras. Paleobotanist, 14:144–171.

1967*a*. Geologic history of the Californian insular flora. *In* Proceedings of the symposium on the biology of the California islands. Santa Barbara Botanic Garden. Pp. 267–315.

1967*b*. Quaternary extinctions of large mammals. Univ. Calif. Publ. Geol. Sci., 74:1–42.

BAILEY, H. P.

1960. A method of determining the warmth and temperateness of climate. Geografiska Annaler, 42(1):1–16.

1964. Toward a unified concept of the temperate climate. Geog. Rev., 54:516–545.

BAILY, J. L.

1950. *Maxwellia,* genus novum of Muricidae. Nautilus, 64(1):9–14.

BAKER, FRED

1902. List of shells collected on San Martin Island, Lower California. Nautilus, 16:40–43.

BANDY, O. L.

1958. Dominant molluscan faunas of the San Pedro basin, California. J. Paleontology, 32:703–714.

BARTSCH, PAUL

1922. A monograph of the American shipworms. U.S. Natl. Mus., Bull. 122, 51 pp.

1931. The west American mollusks of the genus *Acar.* U.S. Natl. Mus., Proc., 80, art. 9:1–4.

BEU, A. G., and P. A. MAXWELL

1968. Molluscan evidence for Tertiary sea temperatures in New Zealand: a reconsideration. Tuatara, 16(1):68–74.

BRODERIP, W. J., and G. B. SOWERBY

1829. Observations on new or interesting Mollusca contained, for the most part, in the Museum of the Zoological Society. Zool. J., London, 4:359–379.

BRONGERSMA-SANDERS, MARGARETHA

1957. Mass mortality in the sea. *In* Treatise on marine ecology and paleoecology, ed. J. W. Hedgpeth. Geol. Soc. Amer., Mem. 67(1):941-1010.

BULLOCK, T. H.

1955. Compensation for temperature in the metabolism and activity of poikilotherms. Biol. Rev., 30:311–342.

BURCH, J. Q., ed.

1944–1946. Distributional list of the west American marine Mollusca from San Diego, California, to the Polar Sea. Minutes of the Conchological Club of Southern California.

CARPENTER, P. P.

1855–1857. Catalogue of the collection of Mazatlan shells in the British Museum collected by Frederick Reigen. London. 555 pp.

1856. Descriptions of new species and varieties of shells, from the California and west Mexican coasts. Zool. Soc. London, Proc., pt. 23 (1855):228–235.

1857. Report on the present state of our knowledge with regard to the Mollusca of the west coast of North America. Brit. Assoc. Adv. Sci., Rept. (1856):159–368.

1864. Supplementary report on the present state of our knowledge with regard to the Mollusca of the west coast of North America. Brit. Assoc. Adv. Sci., Rept. (1863):517–686.

1866. On the Pleistocene fossils collected by Col. E. Jewett at Santa Barbara, California, with descriptions of new species. Ann. Mag. Nat. Hist., ser. 3, 17:274–278.

CARSON, C. M.

1926. New molluscan species from the California Pliocene. Southern Calif. Acad. Sci., Bull. 25:49–62.

CLARK, B. L.

1921. The marine Tertiary of the west coast of the United States: its sequence, paleogeography, and the problems of correlation. J. Geology, 29:583–614.

1929. Stratigraphy and faunal horizons of the Coast Ranges of California. Berkeley, Calif.: privately published. Pp. 1–30, pls. 1–50.

CLARK, W. B., and M. W. TWITCHELL

1915. Mesozoic and Cenozoic Echinodermata of the United States. U.S. Geol. Survey, Mon. 54, 341 pp., 51 pls.

CONRAD, T. A.

1837. Descriptions of marine shells from upper California, collected by Thomas Nuttall, Esq. J. Acad. Nat. Sci. Phila., 7:227–268, pls. 17–20.

1849. Descriptions of two genera and new species of Recent shells. Acad. Nat. Sci. Phila., Proc., 4:121.

1855*a*. Report of Mr. T. A. Conrad on the fossil shells collected in California by Wm. P. Blake. U.S. House of Rep., Doc. no. 129:9–20 [reprinted in Dall, 1909:170].
1855*b*. Descriptions of the fossil shells. U.S. Pacific Railroad Repts., 5, art. 2:318–328.
1856. Descriptions of three new genera, twenty-three new species middle Tertiary fossils from California, and one from Texas. Acad. Nat. Sci. Phila., Proc., 8:312–316.
1867. Descriptions of new west coast shells. Amer. J. Conch., 3:192–193.

Cooper, J. G.
1888. Seventh annual report of the state mineralogist. Calif. State Mining Bureau. Pp. 1–315.

Corey, W. H.
1954. Tertiary basins of southern California. *In* Geology of southern California. Calif. Div. Mines, Bull. 170, chap. 3:73–83.

Cossman, A. E. M.
1903. Essais de paléoconchologie comparée. Paris. Vol. 5.

Crickmay, C. H.
1929. The anomalous stratigraphy of Deadman's Island, California. J. Geology, 37:617–638.

Crìsp, D. J.
1957. Effects of low temperature on the breeding of marine animals. Nature, 179:1138–1139.

Crisp, D. J., and A. J. Southward
1958. The distribution of intertidal organisms along the coasts of the English Channel. J. Mar. Biol. Assoc. U.K., 37:157–208.

Crowell, J. C.
1962. Displacement along the San Andreas fault, California. Geol. Soc. Amer., Spec. Paper 71:1–51.

Cummings, J. C., R. M. Touring, and E. E. Brabb
1962. Geology of the northern Santa Cruz Mountains, California. Calif. Div. Mines, Bull. 181:179–220.

Dall, W. H.
1871. Descriptions of sixty new forms of mollusks from the west coast of America and the North Pacific Ocean, with notes on others already described. Amer. J. Conch., 3:93–160.
1879. Fossil mollusks from later Tertiaries of California. U.S. Natl. Mus., Proc., 1:10–16.
1891. On some new or interesting west American shells obtained from the dredgings of the U.S. Fish Commission Steamer *Albatross* in 1888, and from other sources. U.S. Natl. Mus., Proc., 14:173–191.
1894*a*. On the species of *Mactra* from California. Nautilus, 7:136–138.
1894*b*. Synopsis of the Mactridae of northwest America, south to Panama. Nautilus, 8:39–43.
1896. Note on *Leda caelata* Hinds. Nautilus, 10:70.
1898. Contributions to the Tertiary fauna of Florida. Wagner Free Inst. Sci., Trans., 3, pt. 4:571–947.
1901*a*. Synopsis of the family Cardiidae and of the North American species. U.S. Natl. Mus., Proc., 23:381–392.
1901*b*. Synopsis of the Lucinacea and of the American species. U.S. Natl. Mus., Proc., 23: 779–833.
1902. Synopsis of the family Veneridae and of the North American Recent species. U.S. Natl. Mus., Proc., 26:335–412.
1903. Tertiary fauna of Florida. Wagner Free Inst. Sci., Trans., 3, pt. 6:1219–1654.
1905. Note on *Lucina* (*Miltha*) *childreni* Gray and on a new species from the Gulf of California. Nautilus, 18:110–112.
1909. The Miocene of Astoria and Coos Bay, Oregon. U.S. Geol. Survey, Prof. Paper 59, 276 pp.
1916. Notes on the west American Columbellidae. Nautilus, 30:25–29.
1918. Changes in and additions to molluscan nomenclature. Biol. Soc. Wash., Proc., 31:137.
1919. Descriptions of new species of mollusks of the family Turritidae from the west coast of America and adjacent regions. U.S. Natl. Mus., Proc., 56:1–86.
1921. Summary of the marine shell-bearing mollusks of the northwest coast of America. U.S. Natl. Mus., Bull. 112, 217 pp.

DESHAYES, G. P.

1839. Nouvelles espèces de mollusques, provenant des côtés de la Californie, du Méxique, du Kamtschatka, et de la Nouvelle-Zélande. Révue Zoologique, par la Société Cuvierienne, 2:356–361.

1854. Descriptions of new shells from the collection of Hugh Cuming Esq. Zool. Soc. London, Proc., pt. 22:317–371.

DEVEREUX, IAN

1967. Oxygen isotope paleotemperature measurements on New Zealand Tertiary fossils. New Zealand J. Sci., 10:988–1011.

DICKIE, L. M.

1959. Water temperature and survival of giant scallop. Amer. Fish. Soc., Trans., 88:73.

DURHAM, D. L., and W. O. ADDICOTT

1965. Pancho Rico Formation, Salinas Valley, California. U.S. Geol. Survey, Prof. Paper 524-A:A1–22, pls. 1–5.

DURHAM, J. W.

1948. Age of post–Mint Canyon marine beds. Geol. Soc. Amer., Bull. 59:1386 (abstract).

1949. *Dendraster elsmerensis* Durham, n. sp. Amer. J. Sci., 247:49–62.

1950. Cenozoic marine climates of the Pacific coast. Geol. Soc. Amer., Bull. 61:1243–1264.

1954. The marine Cenozoic of southern California. *In* Geology of southern California. Calif. Div. Mines, Bull. 170, chap. 3:23–31.

EDWARDS, E. C.

1934. Pliocene conglomerates of the Los Angeles basin and their paleoecologic significance. Amer. Assoc. Petrol. Geol., Bull. 18:786–812.

ELDRIDGE, G. H., and RALPH ARNOLD

1907. The Santa Clara Valley, Puente Hills, and Los Angeles oil districts, southern California. U.S. Geol. Survey, Bull. 309, 266 pp.

EMERSON, W. K.

1956. Pleistocene invertebrates from Punta China, Baja California, Mexico. Amer. Mus. Nat. Hist., Bull. 111, art. 4:313–342, pls. 22, 23.

EMERSON, W. K., and E. P. CHACE

1959. Pleistocene mollusks from Tecolote Creek, San Diego, California. San Diego Soc. Nat. Hist., Trans., 12:335–346.

ENGLISH, W. A.

1914. The Fernando Group near Newhall, California. Univ. Calif. Publ. Geol. Sci., 8:203–218.

ESCHSCHOLTZ, F.

1833. Zoologischer Atlas. Berlin: G. Reimer. Vol. 5.

FAUSTMAN, W. F.

1964. Paleontology of the Wildcat Group at Scotia and Centerville Beach, California. Univ. Calif. Publ. Geol. Sci., 41:97–160.

FORBES, E.

1852. On the marine Mollusca discovered during the voyages of the *Herald* and *Pandora*. Zool. Soc. London, Proc., pt. 18 (1850):270–274.

GABB, W. M.

1865. Descriptions of new species of marine shells from the coast of California. Calif. Acad. Sci., Proc., ser. 1, 3:182–190.

1866–1869. Cretaceous and Tertiary fossils. Vol. 2, sec. 1, Palaeontology. Geol. Survey Calif. Pp. 1–124, pls. 1–36.

GALTSOFF, P. S.

1961. Physiology of reproduction in molluscs. Amer. Zool., 1:273–289.

GIESE, A. C.

1959. Comparative physiology: annual reproductive cycles of marine invertebrates. Ann. Rev. Physiology, 21:547–576.

GMELIN, J. F.

1791. Caroli a Linne, Systema naturae per regna tria naturae. Editio decima tertia. London. Vol. 1, pt. 6, cl. 6:3021–3910.

GOULD, A. A.
1846. Descriptions of new species of *Rimula, Crepidula, Calyptraea, Hipponix,* and *Pileopsis,* from the collection of the U.S. Exploring Expedition. Boston Soc. Nat. Hist., Proc., 2:159–162.
1849. Descriptions of shells from the U.S. Exploring Expedition. Boston Soc. Nat. Hist., Proc., 3:89–92.
1850*a*. Descriptions with drawings of several new species of shells from the U.S. Exploring Expedition. Boston Soc. Nat. Hist., Proc., 3:169–172.
1850*b*. Descriptions of shells from the U.S. Exploring Expedition. Boston Soc. Nat. Hist., Proc., 3:214–219.
1853. Descriptions of shells from the Gulf of California and the Pacific coasts of Mexico and California. Boston J. Nat. Hist., 6:374–408.

GRANT, U. S., IV, and H. R. GALE
1931. Catalogue of the marine Pliocene and Pleistocene Mollusca of California and adjacent regions. San Diego Soc. Nat. Hist., Mem. 1, 1036 pp.

GRANT, U. S., IV, and L. G. HERTLEIN
1938. The west American Cenozoic Echinoidea. Univ. Calif. Los Angeles Publ. Math. Phys. Sci., 2, 225 pp.

GRAY, J. E.
1850. On the species of Anomiidae. Zool. Soc. London, Proc., Pt. 17 (1849):113–124.

GUNTER, GORDON
1957. Temperature. *In* Treatise on marine ecology and paleoecology, ed. J. W. Hedgpeth. Vol. 1, Ecology. Geol. Soc. Amer., Mem. 67(1):159–184.

HALL, C. A., JR.
1962. Evolution of the echinoid genus *Astrodapsis*. Univ. Calif. Publ. Geol. Sci., 40:47–180.
1964. Shallow-water marine climates and molluscan provinces. Ecology, 45:226–234.

HANLEY, S.
1844. Description of new species of Mytilacea. Zool. Soc. London, Proc., Pt. 12 (1844):14–17.

HEDGPETH, J. W.
1957. Marine biogeography. *In* Treatise on marine ecology and paleoecology, ed. J. W. Hedgpeth. Vol. 1, Ecology. Geol. Soc. Amer., Mem. 67(1):359–382.

HERTLEIN, L. G.
1928. *Pecten* (*Patinopecten*) *lohri,* new name for *Pecten oweni* Arnold, a Pliocene species from California. Nautilus, 41:93–94.

HERTLEIN, L. G., and A. M. STRONG
1940–1951. Mollusks from the west coast of Mexico and Central America. Zoologica, 25 (1940): 369–430; 28 (1943):149–168; 31 (1946):53–76, 93–120; 32 (1947):129–150; 33 (1948):163–198; 34 (1949):63–97, 239–258; 35 (1950):217–252; 36 (1951):67–120.

HILL, M. L.
1937. Structure of the San Gabriel Mountains north of Los Angeles, California. Univ. Calif. Publ. Geol. Sci., 19:137–163.

HINDS, R. B.
1843*a*. Descriptions of new species of *Nucula,* from the collections of Sir Edward Belcher, C.B., and Hugh Cuming, Esq. Zool. Soc. London, Proc., Pt. 11 (1843):97–101.
1843*b*. Descriptions of new species of *Scalaria* and *Murex,* from the collection of Sir Edward Belcher, C. B. Zool. Soc. London, Proc., Pt. 11 (1843):124–129.
1844. The zoology of the voyage of the H.M.S. *Sulphur,* under the command of Sir Edward Belcher during 1836–1842. London. Pt. 2, no. 7:1–24.

HORNIBROOK, N. DE B.
1967. New Zealand Tertiary microfossil zonation, correlation, and climate. *In* Tertiary correlations and climatic changes in the Pacific (Symposium 25), ed. K. Hatai. Proc. 11th Pac. Sci. Cong. Pp. 29–39.

HUTCHINS, L. W.
1947. The bases for temperature zonation in geographical distribution. Ecol. Mon., 17:325–335.

INGLE, J. C., JR.
1967. Foraminiferal biofacies variation and the Miocene-Pliocene boundary in southern California. Bulls. Amer. Paleontology, 52:209–394.
1968. Pliocene planktonic Foraminifera from northern California and paleo-oceanographic implications (abstract). Geol. Soc. Amer. Program, Annual Meeting, p. 147.

JAHNS, H., and W. R. MUEHLBERGER
1954. Geology of the Soledad basin, Los Angeles County. *In* Geology of southern California, ed. R. H. Jahns. Calif. Div. Mines, Bull. 170, map sheet 6.

JENKINS, D. G.
1968. Planktonic Foraminiferida as indicators of New Zealand Tertiary paleotemperatures. Tuatara, 16:32–37.

JOHNSON, R. G.
1960. Models and methods for analysis of the mode of formation of fossil assemblages. Geol. Soc. Amer., Bull. 71:1075–1086.
1965. Pelecypod death assemblages in Tomales Bay, California. J. Paleontology, 39:80–85.

JORDAN, E. K.
1924. Quaternary and Recent molluscan faunas of the west coast of Lower California. Southern Calif. Acad. Sci., Proc., 23:145–156.

KANAKOFF, G. P.
1956. Two new species of *Nassarius* from the Pliocene of Los Angeles County, California. Southern Calif. Acad. Sci., Bull. 55:110–113.
1966. A new species of *Boetica* from the Pliocene of California. Los Angeles County Mus. Contr. Sci., 103, 4 pp.

KEEN, A. M.
1958. Sea shells of tropical west America. Stanford, Calif.: Stanford Univ. Press. 624 pp.
1962. Nomenclatural notes on some west American mollusks. Veliger, 4:178–180.
1963. Marine molluscan genera of western North America. Stanford, Calif.: Stanford Univ. Press. 126 pp.

KEW, W. S. W.
1924. Geology and oil resources of a part of Los Angeles and Ventura counties, California. U.S. Geol. Survey, Bull. 753, 202 pp.

KEYES, I. W.
1968. Cenozoic marine temperatures indicated by the Scleractinian coral fauna of New Zealand. Tuatara, 16:21–25.

KINNE, O.
1963. The effects of temperature and salinity on marine and brackish water animals. Oceanog. Mar. Biol. Ann. Rev., 1:301–340.
1970. Marine ecology, vol. 1. Environmental factors, pt. 1. New York: Wiley-Interscience. 681 pp.

KORRINGA, P.
1957. Water temperature and breeding throughout the geographical range of *Ostrea edulis*. L'Année Biologique, ser. 3, 3:1–15.

LEWIS, K. B.
1968. Size of fossil animals as an indicator of paleotemperature. Tuatara, 16:62–67.

LIGHT, S. F., R. I. SMITH, F. A. PITELKA, D. P. ABBOTT, and F. M. WEESNER
1964. Intertidal invertebrates of the central California coast. Berkeley and Los Angeles: Univ. Calif. Press. 446 pp.

LINNÉ, C.
1758. Systema naturae per regna tria naturae. Editio decima, reformata. Stockholm. Vol. 1, Regnum animale. 824 pp.

LOOSANOFF, V. L., and C. A. NOMEJKO
1951. Existence of physiologically-different races of oysters, *Crassostrea virginica*. Biol. Bull., 101:151–156.

LUCAS, C. E.
1947. The ecological effects of external metabolites. Biol. Rev., 22:270–295.

1955. External metabolites in the sea. Papers Mar. Biol. Oceanog., Deep Sea Res. Suppl. 3:139–148.

1961. Interrelationships between aquatic organisms mediated by external metabolites. Oceanography, Amer. Assoc. Adv. Sci., Publ. no. 67:499–517.

MacGinitie, G. E.

1955. Distribution and ecology of the marine invertebrates of Point Barrow, Alaska. Smithsonian Misc. Coll., 128:1–201.

McLean, J. H.

1969. Marine shells of southern California. Los Angeles County Mus. Sci. Ser. 24, Zool. no. 11, 98 pp.

Mawe, J.

1823. The Linnaean system of conchology. London: Longman, Hurst, Rees, Orme, and Brown. 207 pp.

Meek, F. B.

1864. Check list of the invertebrate fossils of North America, Miocene. Smithsonian Misc. Coll., 7:1–32.

Melvill, J. C.

1893. Description of a new species of *Mitra*. Conchologist, 2:140–141.

Miller, W. J.

1934. Geology of the western San Gabriel Mountains of California. Univ. Calif. Los Angeles Publ. Math. Phys. Sci., 1:1–114.

Moody, C. L.

1916. Fauna of the Fernando of Los Angeles. Univ. Calif. Publ. Geol. Sci., 10:39–62.

Natland, M. L.

1957. Paleoecology of west coast Tertiary sediments. *In* Treatise on marine ecology and paleoecology, ed. J. W. Hedgpeth. Vol. 2, Paleoecology. Geol. Soc. Amer., Mem. 67(2): 543–571.

Natland, M. L., and P. H. Kuenen

1951. Sedimentary history of the Ventura basin, California, and the action of turbidity currents. Soc. Econ. Paleont. Mineral., Spec. Publ. 2:76–107.

Newell, I. M.

1948. Marine molluscan provinces of western North America: a critique and a new analysis. Amer. Philos. Soc., Proc., 92:155–166.

Nomland, J. O.

1917. The Etchegoin Pliocene of middle California. Univ. Calif. Publ. Geol. Sci., 10: 191–254.

Oakeshott, G. B.

1950. Geology of the Placerita oil field, Los Angeles County, California. Calif. J. Mines Geol., 46:43–80.

1958. Geology and mineral deposits of San Fernando quadrangle, Los Angeles County, California. Calif. Div. Mines, Bull. 172, 147 pp.

Ogle, B. A.

1953. Geology of Eel River Valley area, Humboldt County, California. Calif. Div. Mines, Bull. 164:1–128.

Oldroyd, I. S.

1924. The marine shells of the west coast of North America. Stanford Univ. Publ. Geol. Sci., 1, 248 pp., 57 pls.

1927. The marine shells of the west coast of North America. Stanford Univ. Publ. Geol. Sci., 2, pt. 1, 297 pp.; pt. 2, 304 pp.; pt. 3, 339 pp., 108 pls.

Oldroyd, T. S.

1914. A remarkably rich pocket of fossil drift from the Pleistocene. Nautilus, 28:80–82.

Olsson, A. A.

1961. Mollusks of the tropical eastern Pacific. Panamic-Pacific Pelecypoda. Ithaca, New York: Paleontologic Research Institution. 574 pp., 86 pls.

ORCUTT, C. R.

1886. Notes on the mollusks of the vicinity of San Diego, California, and Todos Santos Bay, Lower California. U.S. Natl. Mus., Proc., 8:534–552.

ORTON, J. H.

1920. Sea-temperature, breeding and distribution in marine animals. J. Mar. Biol. Assoc. U.K., 12:339–366.

PACK, R. W.

1909. Notes on echinoids from the Tertiary of California. Univ. Calif. Publ. Geol. Sci., 5: 275–283.

PACKARD, E. L.

1918. Molluscan fauna from San Francisco Bay. Univ. Calif. Publ. Zool., 14:199–452.

PALMER, K. V. W.

1958. Type specimens of marine Mollusca described by P. P. Carpenter from the west coast. Geol. Soc. Amer., Mem. 76, 376 pp.

PARKER, PIERRE

1949. Fossil and Recent species of the pelecypod genera *Chione* and *Securella* from the Pacific coast. J. Paleontology, 23:577–593.

PILSBRY, H. A., and H. N. LOWE

1933. West Mexican and Central American mollusks collected by H. N. Lowe, 1929–1931. Acad. Nat. Sci. Phila., Proc., 84:33–144.

POTTER, P. E., and F. J. PETTIJOHN

1963. Paleocurrents and basin analysis. New York: Academic Press. 296 pp.

PROSSER, C. L.

1955. Physiological variation in animals. Biol. Rev., 30:229–262.

PYEFINCH, K. A.

1948. Notes on the biology of cirripedes. J. Mar. Biol. Assoc. U.K., 27:464–503.

RAYMONT, J. E. G.

1963. Plankton and productivity in the oceans. Oxford: Pergamon Press. 660 pp.

REDFIELD, J. H.

1846. Description of some new species of shells. Lyceum Nat. Hist. New York, Annals, 4: 163–168.

REED, R. D.

1933*a*. Geology of California. Tulsa: Amer. Assoc. Petrol. Geol. 355 pp.

1933*b*. Oil-bearing Pliocene beds of southern California. Pan-Am Geologist, 59:231 (abstract).

REED, R. D., and J. S. HOLLISTER

1936. Structural evolution of southern California. Tulsa: Amer. Assoc. Petrol. Geol. 157 pp.

REEVE, L.

1843–1878. Conchologia iconica, or illustrations of the shells of molluscous animals. London. Vols. 1–20, with suppl. to *Conus*. Continued by G. B. Sowerby II, beginning with genus *Pyramidella* in vol. 15.

REINHART, P. W.

1937. Three new species of the pelecypod family Arcidae from the Pliocene of California. J. Paleontology, 11:181–185.

REMOND, A.

1863. Description of two new species of *Scutella*. Calif. Acad. Sci., Proc., ser. 1, 3:13–15.

RICKETTS, E. F., and J. CALVIN

1939. Between Pacific tides. Stanford, Calif.: Stanford Univ. Press. 516 pp.

RODEN, G. I.

1964. Oceanographic aspects of Gulf of California. *In* Marine geology of the Gulf of California. Amer. Assoc. Petrol. Geol., Mem. 3:30–58.

SCHENCK, H. G., and A. M. KEEN

1936. Marine molluscan provinces of western North America. Amer. Phil. Soc. Proc., 76: 921–938.

1937. An index-method for comparing molluscan faunules. Amer. Philos. Soc., Proc., 77: 161–182.

1940. California fossils for the field geologist. Stanford, Calif.: Stanford Univ. Press. 88 pp.

SCHWARZBACH, M.

1963. Climates of the past, trans. and ed. R. O. Muir. New York: D. Van Nostrand. 328 pp.

SCRIPPS INSTITUTION OF OCEANOGRAPHY

1960, 1965–1967. Data report, surface water temperatures at shore stations, U.S. west coast. 4 vols.: 1956–1959, 1964, 1965, and 1966.

1965. Oceanic observations of the Pacific. Berkeley and Los Angeles: Univ. of Calif. Press.

SEGAL, E.

1956. Microgeographic variation as thermal acclimation in an intertidal mollusc. Biol. Bull., 111:129–152.

1961. Acclimation in molluscs. Amer. Zool., 1:235–244.

SHEPARD, F. P.

1951. Transportation of sand into deep water. Soc. Econ. Paleont. Mineral., Spec. Publ. 2: 53–65.

SMITH, A. G., and M. GORDON, JR.

1948. The marine mollusks and brachiopods of Monterey Bay, California, and vicinity. Calif. Acad. Sci., Proc., ser. 4, 26:147–245.

SMITH, J. P.

1919. Climatic relations of the Tertiary and Quaternary faunas of the California region. Calif. Acad. Sci., Proc., ser. 4, 9:123–173.

SMITH, J. T.

1970. Taxonomy, distribution, and phylogeny of the cymatiid gastropods *Argobuccinum, Fusitriton, Mediargo,* and *Priene*. Bulls. Amer. Paleontology, 56(254):445–573.

SOWERBY, G. B.

1824. The genera of Recent and fossil shells. London. Vol. 1, 154 pls., pages not numbered.

1825. A catalogue of the shells contained in the collection of the late Earl of Tankerville. London. Pp. 1–92.

1834*a*. Characters of new species of Mollusca and Conchifera. Zool. Soc. London, Proc., Pt. 1 (1833):134–139.

1834*b*. Characters of new species of Mollusca and Conchifera. Zool. Soc. London, Proc., Pt. 2 (1834):87–89.

SOWERBY, G. B., JR.

1847–1887. Thesaurus conchyliorum, or monographs of genera of shells, ed. G. B. Sowerby, Jr., completed by G. B. Sowerby, III. London. Vols. 1–5.

STANTON, R. J.

1966. Megafauna of the upper Miocene Castaic Formation, Los Angeles County, California. J. Paleontology, 40:21–40.

STANTON, R. J., and J. R. DODD

1970. Paleoecologic techniques: comparison of faunal and geochemical analyses of Pliocene paleoenvironments, Kettleman Hills, California. J. Paleontology, 44:1092–1121.

STAUBER, L. A.

1950. Physiological species: oysters and oyster drills. Ecology, 31:109–118.

STEARNS, R. E. C.

1871. Preliminary descriptions of new species of marine Mollusca from the west coast of North America. Conchological Mem. no. 7, separately printed. Petaluma, Calif.: Journal and Argus Press.

1872. Description of a new species of *Mangelia* from California. Calif. Acad. Sci., Proc., ser. 1, 4:226.

1875. Descriptions of new fossil shells from the Tertiary of California. Acad. Nat. Sci. Phila., Proc., 27:463–464.

STEWART, R. B.

1930. Gabb's California Cretaceous and Tertiary type lamellibranchs. Acad. Nat. Sci. Phila., Spec. Publ. 3, 314 pp.

STRONG, A. M., G. D. HANNA, and L. G. HERTLEIN

1933. The Templeton Crocker Expedition of the California Academy of Sciences, 1932. No.

10, Marine Mollusca from Acapulco, Mexico, with notes on other species. Calif. Acad. Sci., Proc., ser. 4, 21:117–130.

SVERDRUP, H. U., M. W. JOHNSON, and R. H. FLEMING
1942. The oceans. New York: Prentice-Hall. 1060 pp.

THORSON, GUNNAR
1946. Reproduction and larval development of Danish marine bottom invertebrates. Medd. Komm. Danmarks Fiskeri-og Havunders., ser. Plankton, 4:1–523.
1957. Bottom communities (sublittoral or shallow shelf). *In* Treatise on marine ecology and paleoecology, ed. J. W. Hedgpeth. Vol. 1, Ecology. Geol. Soc. Amer., Mem. 67(1): 461–521.

TRASK, J. B.
1855. Descriptions of fossil shells. Calif. Acad. Sci., Proc., ser. 1, 1:40–42.

TRASK, P. D.
1922. The Briones Formation of middle California. Univ. Calif. Publ. Geol. Sci., 13:133–174.

TRYON, G. W., JR.
1863. Contributions towards a monograph of the order of Pholadacea, with descriptions of new species. Acad. Nat. Sci. Phila., Proc., 15:143–145.
1879–1913. Manual of conchology. Philadelphia. Ser. 1, vols. 1–17. Continued by H. A. Pilsbry, beginning with vol. 10 (1888), pt. 2:161.

U.S. NAVY
1956. Marine climatic atlas of the world. Vol. 2, North Pacific Ocean. U.S. Govt. Printing Office.

VALENTINE, J. W.
1955. Upwelling and thermally anomalous Pacific coast Pleistocene molluscan faunas. Amer. J. Sci., 253:462–474.
1957. Late Pleistocene faunas from the northwestern coast of Baja California, Mexico. San Diego Soc. Nat. Hist., Trans., 12:289–308.
1960. Habitats and sources of Pleistocene mollusks at Torrey Pines Park, California. Ecology, 41:161–165.
1961. Paleoecologic molluscan geography of the Californian Pleistocene. Univ. Calif. Publ. Geol. Sci., 34:309–442.
1966. Numerical analysis of marine molluscan ranges on the extratropical northeastern Pacific shelf. Limnol. Oceanog., 11:198–211.

VALENTINE, J. W., and R. F. MEADE
1961. Californian Pleistocene paleotemperatures. Univ. Calif. Publ. Geol. Sci., 40:1–46.

VEDDER, J. G.
1960. Previously unreported Pliocene Mollusca from the southeastern Los Angeles basin. *In* Short papers in the geological sciences, art. 151. U.S. Geol. Survey, Prof. Paper 400-B: B326–B328.

VOKES, H. E.
1969. Observations on the genus *Miltha* (Mollusca: Bivalvia) with notes on the type and the Florida Neogene species. Tulane Studies in Geology and Paleontology, 7:93–126.

WATTS, W. L.
1901. Oil- and gas-yielding formations of California. Calif. Div. Mines, Bull. 19:56.

WEAVER, C. E., et al.
1944. Correlation of the marine Cenozoic formations of western North America. Geol. Soc. Amer., Bull. 55:569–598.

WHITNEY, J. D.
1865. Geology of California. Vol. 1, Geology. Geol. Survey Calif. 498 pp.

WILLET, G.
1943. Northeast American species of *Glycymeris*. Southern Calif. Acad. Sci., Bull. 42:107–114.

WILLIS, ROBIN

1952. Placerita oil field. Amer. Assoc. Petrol. Geol., Soc. Econ. Paleont. Mineral., Soc. Econ. Geol. Guidebook. Los Angeles. Pp. 32–41.

WINTERER, E. L., and D. L. DURHAM

1951. A formation of late Miocene and early Pliocene age on the north slope of the Santa Susana Mountains, California. Amer. Assoc. Petrol. Geol., Bull. 35:2631 (abstract).

1954. Geology of a part of the eastern Ventura basin, Los Angeles County. Calif. Div. Mines, Bull. 170, map sheet 5.

1962. Geology of southeastern Ventura basin, Los Angeles County, California. U.S. Geol. Survey, Prof. Paper 334-H:275–366.

WOLFE, J. A., and D. M. HOPKINS

1967. Climatic changes recorded by Tertiary land floras of North-Western North America. *In* Tertiary correlations and climatic changes in the Pacific (Symposium 25), ed. K. Hatai. Proc. 11th Pac. Sci. Cong. Pp. 67–76.

WOODFORD, A. O., J. E. SCHOELLHAMER, J. G. VEDDER, and R. F. YERKES

1954. Geology of the Los Angeles basin. *In* Geology of southern California. Calif. Div. Mines, Bull. 170, chap. 2:65–81.

WOODRING, W. P.

1938. Lower Pliocene mollusks and echinoids from the Los Angeles basin, California. U.S. Geol. Survey, Prof. Paper 190, 67 pp.

1951. Basic assumption underlying paleoecology. Science, 113:482–483 (abstract).

WOODRING, W. P., and M. N. BRAMLETTE

1950. Geology and paleontology of the Santa Maria district, California. U.S. Geol. Survey, Prof. Paper 222, 185 pp.

WOODRING, W. P., M. N. BRAMLETTE, and W. S. W. KEW

1946. Geology and paleontology of Palos Verdes Hills, California. U.S. Geol. Survey, Prof. Paper 207, 114 pp.

WOODRING, W. P., R. STEWART, and R. W. RICHARDS

1940. Geology of Kettleman Hills oil field, California. U.S. Geol. Survey, Prof. Paper 195, 170 pp.